AF443425

STEAM-H: Science, Technology, Engineering, Agriculture, Mathematics & Health

STEAM-H: Science, Technology, Engineering, Agriculture, Mathematics & Health

Series Editor
Bourama Toni
Department of Mathematics
Howard University
Washington, DC, USA

This interdisciplinary series highlights the wealth of recent advances in the pure and applied sciences made by researchers collaborating between fields where mathematics is a core focus. As we continue to make fundamental advances in various scientific disciplines, the most powerful applications will increasingly be revealed by an interdisciplinary approach. This series serves as a catalyst for these researchers to develop novel applications of, and approaches to, the mathematical sciences. As such, we expect this series to become a national and international reference in STEAM-H education and research.

Interdisciplinary by design, the series focuses largely on scientists and mathematicians developing novel methodologies and research techniques that have benefits beyond a single community. This approach seeks to connect researchers from across the globe, united in the common language of the mathematical sciences. Thus, volumes in this series are suitable for both students and researchers in a variety of interdisciplinary fields, such as: mathematics as it applies to engineering; physical chemistry and material sciences; environmental, health, behavioral and life sciences; nanotechnology and robotics; computational and data sciences; signal/image processing and machine learning; finance, economics, operations research, and game theory.

The series originated from the weekly yearlong STEAM-H Lecture series at Virginia State University featuring world-class experts in a dynamic forum. Contributions reflected the most recent advances in scientific knowledge and were delivered in a standardized, self-contained and pedagogically-oriented manner to a multidisciplinary audience of faculty and students with the objective of fostering student interest and participation in the STEAM-H disciplines as well as fostering interdisciplinary collaborative research. The series strongly advocates multidisciplinary collaboration with the goal to generate new interdisciplinary holistic approaches, instruments and models, including new knowledge, and to transcend scientific boundaries.

More information about this series at http://www.springer.com/series/15560

Norou Diawara

Editor

Modern Statistical Methods for Spatial and Multivariate Data

 Springer

Editor
Norou Diawara
Department of Mathematics and Statistics
Old Dominion University
Norfolk, VA, USA

ISSN 2520-193X ISSN 2520-1948 (electronic)
STEAM-H: Science, Technology, Engineering, Agriculture, Mathematics & Health
ISBN 978-3-030-11430-5 ISBN 978-3-030-11431-2 (eBook)
https://doi.org/10.1007/978-3-030-11431-2

Mathematics Subject Classification: 62Hxx, 62H11, 30C40

Preface

Statistical ideas and concepts have increasing impacts at many levels. From H.G. Wells' 1903 book *Mankind in the Making*, a quote paraphrased by Sam Wilks in his 1950 American Statistical Association speech states: "Statistical thinking will one day be as necessary for efficient citizenship as the ability to read and write." Current disciplinary boundaries encourage interaction between scientists and the sharing of information and educational resources. Researchers from these interdisciplinary fields will find this book an important resource for the latest statistical methods for spatial and multivariate data.

Given the increasingly complex data world we live in, this volume takes on unique approach with respect to methodology, simulation, and analysis. The environment provides the perfect setting for exciting opportunities and interdisciplinary research and for practical and robust solutions, contributing to the science community in large. The National Institutes of Health and the Howard Hughes Medical Institute have strongly recommended that undergraduate biology education should incorporate mathematics and statistics, physics, chemistry, computer science, and engineering until "interdisciplinary thinking and work become second nature." In that sense, this volume is playing an ever more important role in the physical and life sciences, engineering and technology, data sciences, and artificial intelligence, blurring the boundaries between scientific disciplines.

The shared emphasis of these carefully selected and refereed contributed chapters is on important methods, research directions, and applications of analysis including within and beyond mathematics and statistics. As such the volume promotes statistical sciences and their applications to physical, life, and data sciences. Statistical methods for spatial and multivariate data have gained indeed tremendous interest over the last decades and are rapidly expanding. This book features recent advances in statistics to include the spatio-temporal aspects, classification techniques, the multivariate outcomes with zero and doubly inflated data, the copula distributions, the wavelet kernels for support matrix machines, and feasible algorithmic solutions. Events are sometimes affected by a set of covariates accounted in space locations and times. With the influx of big data, statistical tools are identified, tested, and improved to fill in the gaps sometimes found in the environmental, financial, and healthcare fields.

This volume stretches our boundaries of knowledge for this fascinating and ongoing area of research. It features the following chapters:

The chapter "Functional Form of Markovian Attribute-Level Best-Worst Discrete Choice Modelling" by Amanda Working, Mohammed Alqawba, and Norou Diawara provides modeling discrete choice experiments. The challenging parts can be linked to the large number of covariates, issues with reliability, and the condition that consumer behaviors is a forward evolving activity/practice. By extending the idea of stationary process, the authors present a dynamic model with evaluation under random utility analysis. The simulated and aggregated data examples show the flexibility and wide applications of the proposed techniques.

The chapter "Spatial and Spatio-temporal Analysis of Precipitation Data from South Carolina" from David Hitchcock, Haigang Liu, and S. Zahra Samadi presents both spatial and spatio-temporal models for rainfall in South Carolina during a period including one of the most destructive storms in state history. The models proposed have allowed to determine several covariates that affect the rainfall and to interpret their effects.

The chapter "A Sparse Areal Mixed Model for Multivariate Outcomes, with an Application to Zero-Inflated Census Data" from Donald Musgrove, Derek S. Young, John Hughes, and Lynn E. Eberly describes the multivariate sparse areal mixed model (MSAMM) as an alternative to the multivariate conditional autoregressive (MCAR) models. The MSAMM is capable of providing superior fit relative to models provided under independent or univariate assumptions.

The next chapter "Wavelet Kernels for Support Matrix Machines" by Edgard M. Maboudou-Tchao provides support vector machine techniques to the matrix-based method support matrix machines (SMM), accepting matrix as input, and then proposing new wavelet kernels for SMM. Such techniques are very powerful approximations for nonstationary signals.

In the chapter "Properties of the Number of Iterations of a Feasible Solutions Algorithm," the authors Sarah A. Janse and Katherine L. Thompson provide statistical guidance for the number of iterations by deriving a lower bound on the probability of obtaining the statistically optimal model in a number of iterations of algorithm along with the performances of the bound.

Classification techniques are commonly used by scientists and businesses alike for decision-making. They involve assignment of objects (or information) to pre-defined groups (or classes) using certain known characteristics such as classifying emails as real or spam using information in the subject field. In the chapter "A Primer of Statistical Methods for Classification," the authors Rajarshi Dey and Madhuri S. Mulekar describe two soft and four hard classifiers popularly used by statisticians in practice. To demonstrate their applications, two simulated and three real-life datasets are used to develop classification criteria. The results of different classifiers are compared using misclassification rate and an uncertainty measure.

In the chapter entitled "A Doubly-Inflated Poisson Distribution and Regression Model" by Manasi Sheth-Chandra, N. Rao Chaganty, and R. T. Sabo, doubly inflated Poisson distribution is presented to account for count inflation at some value k in addition to that seen at zero, while it was also incorporated into the generalized linear models framework to account for associations with covariates.

The chapter "Multivariate Doubly Inflated Negative Binomial Distribution Using Gaussian Copula" by Joseph Mathews, Sumen Sen, and Ishapathik Das presents a model for doubly inflated count data using the negative binomial distribution, under Gaussian copula methods. The authors also provide visuals of the bivariate doubly inflated negative binomial model.

Moran's Index is a statistic that measures spatial autocorrelation, quantifying the degree of dispersion (or spread and properties) of components in some location/area. Recognizing that a single Moran's statistic may not give a sufficient summary of the spatial autocorrelation measure, local spatial statistics have been gaining popularity. Accordingly, the chapter "Quantifying Spatio-temporal Characteristics via Moran's Statistics" by Jennifer Matthews, Norou Diawara, and Lance Waller proposes to partition the area and compute the Moran's statistic of each subarea.

The book as a whole certainly enhances the overall objective of the series, that is, to foster the readership interest and enthusiasm in the STEAM-H disciplines (Science, Technology, Engineering, Agriculture, Mathematics, and Health), to include statistical and data sciences, and to stimulate graduate and undergraduate research through effective interdisciplinary collaboration.

The editor of the current volume is affiliated with the Department of Mathematics and Statistics at Old Dominion University, Norfolk, Virginia. The department has the unique distinction of being the only one in the Commonwealth of Virginia Hampton area to offer B.S., M.S., and Ph.D. degrees in Computational and Applied Mathematics, with an active research program supported by NASA, NSF, EVMS, JLab, and the Commonwealth of Virginia.

Washington, DC, USA   Bourama Toni
Norfolk, VA, USA   Norou Diawara

Acknowledgments

We express special gratitude to the STEAM-H series Editor-in-Chief, Bourama Toni, Professor and Chair, in the Department of Mathematics at Howard University, for the opportunity to produce such a volume. Among the many we thank and acknowledge are the editors and associate editor at Springer, in particular Ms. Dimana Tzvetkova. Your inputs have facilitated and made the presentation very elegant. We are grateful to the many reviewers, both internal and external, for their professionalism; they raised many questions and challenged the materials featured in this book with insightful comments and suggestions from concepts to exposition.

We would like to express our most sincere appreciation to all the contributors for the excellent research work. They all made this volume a reality for the greater benefit of the community of Statistical Scientists.

Contents

Contributors

Mohammed Alqawba Department of Mathematics and Statistics, Old Dominion University, Norfolk, VA, USA

N. Rao Chaganty Department of Mathematics and Statistics, Old Dominion University, Norfolk, VA, USA

Ishapathik Das Department of Mathematics, Indian Institute of Technology Tirupati, Tirupati, India

Rajarshi Dey Department of Mathematics and Statistics, University of South Alabama, Mobile, AL, USA

Norou Diawara Department of Mathematics and Statistics, Old Dominion University, Norfolk, VA, USA

Lynn E. Eberly Division of Biostatistics, University of Minnesota, Minneapolis, MN, USA

David B. Hitchcock Department of Statistics, University of South Carolina, Columbia, SC, USA

John Hughes Department of Biostatistics and Informatics, University of Colorado, Denver, CO, USA

Sarah A. Janse Center for Biostatistics, The Ohio State University, Columbus, OH, USA

Haigang Liu Department of Statistics, University of South Carolina, Columbia, SC, USA

Jennifer L. Matthews Department of Mathematics and Statistics, Old Dominion University, Norfolk, VA, USA

Edgard M. Maboudou-Tchao Department of Statistics, University of Central Florida, Orlando, FL, USA

Joseph Mathews Department of Mathematics and Statistics, Austin Peay State University, Clarksville, TN, USA

Madhuri S. Mulekar Department of Mathematics and Statistics, University of South Alabama, Mobile, AL, USA

Donald Musgrove Medtronic, Minneapolis, MN, USA

Roy T. Sabo Department of Biostatistics, Virginia Commonwealth University, Richmond, VA, USA

S. Zahra Samadi Department of Civil and Environmental Engineering, University of South Carolina, Columbia, SC, USA

Sumen Sen Department of Mathematics and Statistics, Austin Peay State University, Clarksville, TN, USA

Manasi Sheth-Chandra Center for Global Health, Old Dominion University, Norfolk, VA, USA

Katherine L. Thompson Department of Statistics, University of Kentucky, Lexington, KY, USA

Lance A. Waller Department of Biostatistics, Rollins School of Public Health, Emory University, Atlanta, GA, USA

Amanda Working Department of Mathematics and Statistics, Old Dominion University, Norfolk, VA, USA

Derek S. Young Department of Statistics, University of Kentucky, Lexington, KY, USA

Functional Form of Markovian Attribute-Level Best-Worst Discrete Choice Modeling

Amanda Working, Mohammed Alqawba, and Norou Diawara

1 Introduction and Motivation

In today's consumer society, people are provided multiple alternatives in product from which to choose the one that benefits them the most. Such situations include purchasing a smartphone, choosing a car insurance company, or deciding upon vacation destinations. Discrete choice experiments (DCEs) are designed to elicit information from consumers as to why they choose the products or services that they do. DCEs are applicable in multiple fields such as health system program, public policy, transportation research, marketing, and economics.

Discrete choice models (DCMs) are the statistical models describing individuals' preferences for products or services. In traditional DCEs, individuals are given a series of hypothetical scenarios described by attributes forming what is called a *profile*. The set of profiles is the choice set, from which the respondent chooses the best alternative that suits their needs the most as proposed in Louviere et al. (2000). Note that traditional DCEs and their models are built around the work done by Thurstone (1927) and the theoretical basis for it by McFadden (1974).

Although some information are gained from traditional DCEs, they fail to provide knowledge about the impact of the attributes when comparing the utilities between alternatives (Flynn et al. 2007). However, best-worst scaling (BWS) experiments (Marley and Louviere 2005) addressed such issues by asking the respondents to choose the best and worst alternatives from a choice set instead of

A. Working · M. Alqawba · N. Diawara (✉)
Department of Mathematics and Statistics, Old Dominion University, Norfolk, VA, USA
e-mail: ndiawara@odu.edu

© Springer Nature Switzerland AG 2019
N. Diawara (ed.), *Modern Statistical Methods for Spatial and Multivariate Data*,
STEAM-H: Science, Technology, Engineering, Agriculture, Mathematics & Health,
https://doi.org/10.1007/978-3-030-11431-2_1

just the best as is done in the traditional DCEs. Besides having the information of what the best and worst are for the respondents, Marley and Louviere (2005) stated that BWS experiments provide information about the respondent's ranking of products.

BWS experiments are divided into three cases: the object case, the profile case, and the multi-profile case (Louviere et al. 2015). In an object case, a list of objects, scenarios, or attributes are given to respondents and the latter choose the best and worst alternative. Unlike in traditional DCEs, no information about the object is provided to the respondents. In the profile case which is also known as attribute-level best-worst case, information or attributes about the alternatives are provided. In this type of experiments, profiles composed of attribute-levels for each attribute describing a product are determined. From the profiles, respondents are tasked with choosing the best and the worst attribute-level pair. These experiments seek to determine the extent to which attribute and their associated attribute-levels affect consumer behavior. Furthermore, in attribute-level best-worst DCEs the levels of the attributes are well defined and vary across profiles or products, providing sufficient information to measure their impact. For example, Knox et al. (2012) study women's contraceptives by using unbalanced design, with seven attributes (product, effect on acne, effect on weight, frequency of administration, contraceptive effectiveness, effect on bleeding, and cost) with associated number of attribute-levels 8, 3, 4, 4, 8, 9, and 6, respectively. The attribute effect on acne had three levels (worsens, improves, or no effect). Finally, the multi-profile case most closely resembles the traditional DCEs in such that a set of profiles describing products are provided to the respondents and the respondents choose the best and worst products from the choice set.

In this chapter, extension to the existing work done on partial profile models for attribute-level best-worst DCEs, or profile-case BWS, is presented. Attribute-level best-worst data are presented as indicator functions demonstrating the equivalence of these models to the traditional attribute-level best-worst models. The indicator functions are then generalized providing an alternative method for accounting for the attributes of attribute-level models. The functional form of the data definition provides an adaptive model able to conform changes in the profile or set of attributes under discrete choice modeling (DCM). We also allow changes in decisions/utilities over time under Markov decision process (MDP). The conditional logit model is used in the DCMs.

The chapter is organized as follows: attribute-level best-worst DCMs are introduced in Sect. 2. In Sect. 3, functional form of attribute-level best-worst DCMs is presented. Section 4 considers Markov decision process (MDPs) with regard to time sensitive attribute-level best-worst DCEs. Simulated data example of functional form of Markovian attribute-level best-worst DCMs over time and results are described in Sect. 5. We end with a conclusion in Sect. 6.

2 Attribute-Level Best-Worst Discrete Choice Model

2.1 *Literature Review*

Discrete choice experiments (DCEs) and their modeling describe consumer behaviors. Given a set of descriptors about a product, one can estimate the probability an alternative is chosen provided a statistical model appropriate to the data. However, these models are limited in the information they provide. According to Lancsar et al. (2013), there exist only two ways to gain more information from traditional DCEs: either increase the sample size or increase the number of choice sets evaluated by respondents with the risk of adding burden on the respondents in the experiments. Alternatively, Louviere and Woodworth (1991) and Finn and Louviere (1992) presented best-worst scaling experiments that are modified DCEs designed to elicit more information about choice behavior than the pick one approach implemented in the traditional DCEs without the added burden on the respondents.

Although the experiments were presented in the early 1990s, it was not until Marley and Louviere (2005) that the mathematical probabilities and properties were formally presented. Marley and Louviere (2005), Marley et al. (2008), and Marley and Pihlens (2012) provided the probability and properties to best-worst scaling experiments for the profile case and in multi-profile version; however, utility was not introduced. Additionally, Lancsar et al. (2013) provided the probability and utility definition for the multi-profile experiments that include the sequential best-worst choice from a set of choices. Louviere and Woodworth (1983) stated that orthogonal main effects and fractional factorial designs provide better parameter estimates than other designs. In application to best-worst scaling experiments, balanced incomplete block designs (BIBD) (Louviere et al. 2013; Parvin et al. 2016) and orthogonal main effects plans (OMEPs) are popular designs (Flynn et al. 2007; Knox et al. 2012; Street and Knox 2012). These designs and their properties are examined by Street and Burgess (2007). Louviere et al. (2013) looked at the design of experiments for best-worst scaling experiments and stated that it is possible to determine individual parameter estimates for the respondents.

This chapter focuses on the profile case, also known as attribute-level best-worst DCEs. These experiments seek to determine the extent to which attributes and their associated attribute-levels impact consumer behavior. Louviere and Timmermans (1990) introduced hierarchical information integration (HII) for the examination of the valuation of attributes in DCEs. Under HII, the impact of an attribute necessitates discerning the various levels of the attribute. An experiment must be designed in such a way that can measure the different levels varying across products and determine such an impact. In attribute-level best-worst DCEs, the levels of attributes are well defined and vary across profiles, or products, providing sufficient information to measure their impact. Attribute-level best-worst discrete choice experiments provide more information into consumer's choices of products than the usual discrete choice experiments and add more value to the understanding of the data (Marley and Louviere 2005). Those models outperform the standard logit modeling in terms of goodness of fit as mentioned in Hole (2011) in the context of attribute attendance.

Understanding the impact attribute and attribute-levels have on utility is desirable. The guiding ideology in DCEs is that consumers behave in a way to maximize utility. Understanding the impacts attributes and attribute-levels have on consumer behavior provides information with regard to developing and advertising a product, service, or policy to consumers. A preponderance of the literature on attribute-level best-worst DCEs are empirical studies often in the area of health systems research and marketing. Examples include Flynn et al. (2007) on seniors' quality of life, Coast et al. (2006) and Flynn et al. (2008) on dermatologist consultations, Marley and Pihlens (2012) on cell phones, Knox et al. (2012, 2013) on choices in contraceptives for women.

While there exists literature on attribute-level best-worst DCEs, it is rather scarce compared to the work done on traditional DCEs. In this section, we provide utility definition and the resulting choice probabilities and properties. We use the utility definition and choice probabilities to extend the work done by Grossmann et al. (2009) to fit models on a function of the attributes and attribute-levels to reflect fluctuation that are inherent in DCE over time.

2.2 Notations, Theory, and Properties

Attribute-level best-worst scaling are modified DCEs designed to elicit the impact the attributes and attribute-levels have on the utility of a product. As mentioned by Louviere and Timmermans (1990), an experiment must be designed in a way to evaluate combinations of attribute-levels to obtain information about attribute impacts on utility. Best-worst attribute-level DCEs provide such an experimental design to attain these impacts.

In the attribute-level best-worst DCEs, each product is represented by a profile $\mathbf{x} = (x_1, x_2, \ldots, x_K)$, where x_k is the attribute-level for the kth attribute A_k that makes up the product with $k = 1, 2, \ldots, K$. The attribute-levels take values from 1 to l_k for $k = 1, 2, \ldots, K$. The number of possible profiles is given by $\Pi_{k=l}^{K} l_i$. Full factorial designs are generally not used due to the large number of profiles. Alternatively, OMEP designs are promoted in the literature as efficient and optimal provide sufficient information to estimate model parameters (Louviere and Woodworth 1983; Street and Burgess 2007; Street and Knox 2012).

In these experiments, respondents are tasked with choosing a pair of attribute-levels that contains the best and the worst attribute-level for a given profile. For every profile, the choice set is then:

$$C_x = \{(x_1, x_2), \ldots, (x_1, x_K), (x_2, x_3), \ldots, (x_{K-1}, x_K), (x_2, x_1), \ldots, (x_K, x_{K-1})\},$$

where the first attribute-level is considered to be the best and the second is the worst. From the profile C_x, the respondent determines from the $\tau = K(K-1)$ choices given which is the best-worst pair.

In our setup, we extend the state of choices as follows. Let there be G profiles and the associated profiles given as:

$$x_1 = (x_{11}, x_{12}, \ldots, x_{1K})$$

$$x_2 = (x_{21}, x_{22}, \ldots, x_{2K})$$

$$\vdots$$

$$x_G = (x_{G1}, x_{G2}, \ldots, x_{GK}).$$

The corresponding choice sets for the G profiles are given in Fig. 1. To simplify the notation, we may interchange C_1 with C_{x_1}, C_2 with $C_{x_2} \ldots$, and C_G with C_{x_G}.

The total number of attribute-levels is $L = \sum_{i=1}^{k} l_i$, and $J = \sum_{k=1}^{K} l_k(L - l_k)$ is the total number of unique attribute-level pairs in the experiment (Street and Knox 2012). Within each of the G choice sets, there are $\tau = K(K-1)$ choice pairs.

In the experiment, there is a total of $J = \sum_{k=1}^{K} l_k(L - l_k)$ alternatives. However, within a choice set there is a total of $\tau = K(K-1)$ choices in each of the G choice sets evaluated. Each respondent will have made G choices within the experiment. The response variable representing the choices within each of the choice sets for the

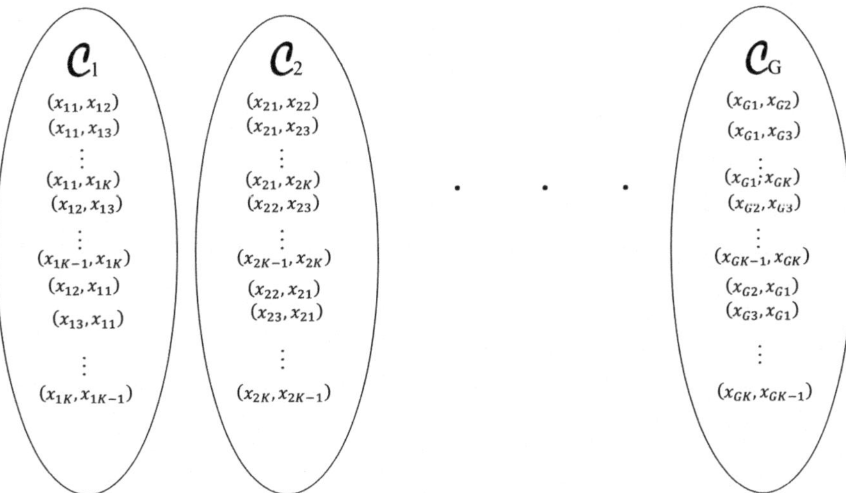

Fig. 1 The G choice sets in an experiment with corresponding choice pairs

experiment is binary data and denoted as:

$$
Y_{isj} = \begin{cases} 1, & \text{if } s\text{th respondent chooses } j\text{th alternative in the } i\text{th choice set,} \\ 0, & \text{otherwise,} \end{cases} \tag{2.1}
$$

for $i = 1, 2, \ldots, G$, $\quad s = 1, 2, \ldots, n \quad$ and $\quad j = 1, 2, \ldots, \tau$.

For the attribute-level best-worst DCEs, the data, $\mathbf{X}$, is composed of indicators for the best and worst attributes and attribute-levels. Consider the choice pair $(x_{ij}, x_{ij'})$ from the choice set $\mathcal{C}_i$, for $i = 1, 2, \ldots, G$, $j \neq j'$, $j, j' = 1, 2, \ldots, K$, and $1 \leq x_{ij} \leq l_j$. Let $\mathbf{X}$ be the $J \times p$ design matrix, where $p = K + \sum_{k=1}^{K} l_k$. The rows of $\mathbf{X}$ correspond to the possible choice pairs. Let $X_{A_1}, X_{A_2}, \ldots, X_{A_K}$ be the data corresponding to the attributes $A_k, k = 1, 2, \ldots, K$. Then,

$$
X_{A_k} = \begin{cases} 1, & \text{if } x_{ij} \in A_k \text{ for } k = 1, 2, \ldots, K, \\ -1, & \text{if } x_{ij'} \in A_k \text{ for } k = 1, 2, \ldots, K, \\ 0, & \text{otherwise.} \end{cases}
$$

Let $X_{A_k x_{ik}}$ be the data for the attribute-level $1 \leq x_{ik} \leq l_k$ within attribute A_k, $\quad \forall k = 1, 2, \ldots, K$. Referring to the choice pair $(x_{ij}, x_{ij'})$, the corresponding data for the attribute-levels are given by,

$$
X_{A_k x_{ik}} = \begin{cases} 1, & \text{if } x_{ij} = x_{ik} \in A_k \text{ is the best attribute-level,} \\ -1, & \text{if } x_{ij'} = x_{ik} \in A_k \text{ is the worst attribute-level,} \\ 0, & \text{otherwise.} \end{cases}
$$

Marley and Louviere (2005) developed the probability theory for best-worst scaling experiments including attribute-level best-worst DCEs. In attribute-level best-worst DCEs, there are two components being modeled, the best choice and the worst choice of attribute-levels from a profile $\mathbf{x}_i$, where $i = 1, 2, \ldots, G$. Under random utility theory (Marschak 1960), there are random utilities U_{ij} corresponding to the τ attribute-levels in the choice set and an individual chooses an alternative with highest utility, i.e., they are not independent. Consider the choice pair $(x_{ij}, x_{ij'})$, for $i = 1, 2, \ldots, G$, $j, j' = 1, 2, \ldots, K$, and $j \neq j'$. According to Marley and Louviere (2005), the definition of utility consistent with random utility theory satisfies, $U_{ij} = -U_{ij'}$ and $U_{ijj'} = U_{ij} - U_{ij'}$ for $i = 1, 2, \ldots, G$, $j, j' = 1, 2, \ldots, K$, and $j \neq j'$, and $U_{ij} = V_{ij} + \epsilon_{ij}$ where V_{ij} is a systematic component and ϵ_{ij} is an error term that distributed as type I extreme value distribution (McFadden 1978). Hence, the definition of utility associated with the best-worst choice pair under random utility theory is given by:

$$
U_{ijj'} = U_{ij} - U_{ij'} = V_{ij} - V_{ij'} + \epsilon_{ij} - \epsilon_{ij'} \tag{2.2}
$$

for $i = 1, 2, \ldots, G, j, j' = 1, 2, \ldots, K$, and $j \neq j'$.

The definition of the utilities under the random utility model is unable to be modeled under the conditional logit model due to the definition of the error components (Marley and Louviere 2005).

If we assume that the random error terms are independently and identically distributed type I extreme value distribution, or the Gumbel distribution, then the choice probability comes directly from the conditional logit. The choice probability is then,

$$
\begin{aligned}
BW_{x_i}(x_{ij}, x_{ij'}) &= P(U_{ijj'} > U_{iqq'}, \forall q, q' \in \mathcal{C}_i) \\
&= P(V_{ijj'} + \epsilon_{ijj'} > V_{iqq'} + \epsilon_{iqq'}, \forall q, q' \in \mathcal{C}_i) \\
&= P(\epsilon_{iqq'} - \epsilon_{ijj'} < V_{ijj'} + V_{iqq'}, \forall q, q' \in \mathcal{C}_i),
\end{aligned}
\tag{2.3}
$$

where $j \neq j'$, $j, j' = 1, 2, \ldots, K$, and $i = 1, 2, \ldots, G$.

Since the error terms come from the type I extreme value distribution, their difference is a logistic distribution. It follows from McFadden (1974) that the best-worst attribute-level choice probability is defined by the conditional logit as:

$$
BW_{x_i}(x_{ij}, x_{ij'}) = \frac{exp(V_{ijj'})}{\displaystyle\sum_{(x_{iq}, x_{iq'}) \in \mathcal{C}_i} exp(V_{iqq'})},
\tag{2.4}
$$

where $j \neq j'$, $j, j' = 1, 2, \ldots, K$, and $i = 1, 2, \ldots, G$.

Marley et al. (2008) provide essential properties to the above probabilities. They define the choice probability as:

$$
BW_{\mathbf{x}_i}(x_{ij}, x_{ij'}) = \frac{\dfrac{b(x_{ij})}{b(x_{ij'})}}{\displaystyle\sum_{\forall(x_{ij}, x_{ij'}) \in \mathcal{C}_{x_i}, j \neq j'} \dfrac{b(x_{ij})}{b(x_{ij'})}},
\tag{2.5}
$$

where x_{ij} is chosen as the best attribute-level, and $x_{ij'}$ is the worst attribute-level, and b is some positive scale function or impact of attribute for $j \neq j'$, $j, j' = 1, 2, \ldots, K$, and $i = 1, 2, \ldots, G$. Under the conditional logit, the scale function is defined as $b(x_{ij}) = exp(V_{ij})$, and the probability is as given in Eq. (2.4).

Essential properties of probability hold for Eq. (2.5), as

$$
BW_{x_i}(x_{ij}, x_{ij'}) \geq 0, \quad \forall i, j
\tag{2.6}
$$

and

$$
\sum_{\forall(x_{ij}, x_{ij'}) \in \mathcal{C}_{x_i}, j \neq j'} BW_{x_i}(x_i, x_j) = 1,
\tag{2.7}
$$

where $j, j' = 1, 2, \ldots, K$, $j \neq j'$, and $\forall i = 1, 2, \ldots, G$.

With such assumptions, the consumer is expected to select choices with higher BW_{x_i} values. We denote $BW_{x_i}(x_{ij}, x_{ij'})$ as $P^i_{jj'}$. Attribute-level best-worst models are called maxdiff models because they maximize the difference in utility.

Associated properties of the maxdiff model mentioned in Marley et al. (2008) are:

1. **Invertibility:** For profile **i**,

$$P^i_{jj'} P^i_{j'j} = P^i_{qq'} P^i_{q'q},$$

   where $1 \leq j, j', q, q' \leq k$ and $j \neq j'$ and $q \neq q'$.
2. **Reversibility:** For profile **i** and **i$'$**

$$P^i_{jj'} P^{i'}_{qq'} = P^{i'}_{q'q} P^i_{j'j},$$

   where $x_{ij'} = x_{i'q}$, and $j \neq j'$ and $q \neq q'$.
3. **Reversibility:** For profiles **i**, **i$'$**, and **i$''$**,

$$P^i_{jj'} P^{i'}_{qq'} P^{i''}_{rr'} = P^{i''}_{r'r} P^{i'}_{q'q} P^i_{j'j},$$

   where $x_{ij'} = x_{i'q}$, $x_{i'q'} = x_{i''r}$, and $x_{ij} = x_{i''r'}$, and $j \neq j'$, $q \neq q'$, and $r \neq r'$.
4. **Reversibility:** For profiles **i**, **i$'$**, **i$''$**, and **i$'''$**,

$$P^i_{jj'} P^{i'}_{qq'} P^{i''}_{rr'} P^{i'''}_{ww'} = P^{i'''}_{w'w} P^{i''}_{r'r} P^{i'}_{q'q} P^i_{j'j},$$

   where $x_{ij'} = x_{i'q}$, $x_{i'q'} = x_{i''r}$, $x_{i''r} = x_{i'''w}$, and $x_{ij} = x_{i'''w'}$, and $j \neq j'$, $q \neq q'$, $r \neq r'$ and $w \neq w'$.

Now, the systematic component in Eq. (2.2) can be expressed as,

$$V_{ijj'} = V_{ij} - V_{ij'} = (\mathbf{x}_{ij} - \mathbf{x}_{ij'})' \boldsymbol{\beta}, \tag{2.8}$$

where $j \neq j'$, $j, j' = 1, 2, \ldots, K$, and $i = 1, 2, \ldots, G$. The data x_{ij} are indicators of the attribute $x_{ij} \in A_j$ and the attribute-level x_{ij}. The systematic component V_{ij} can be written as,

$$V_{ij} = \mathbf{x}'_{ij} \boldsymbol{\beta} = \beta_{A_j} + \beta_{x_{ij} A_j}, \tag{2.9}$$

where $j \neq j'$, $j, j' = 1, 2, \ldots, K$, and $i = 1, 2, \ldots, G$.

We assume the error terms come from a type I extreme value distribution and use the conditional logit model to estimate the $p \times 1$ parameter vector,

$$\beta' = (\beta_{A_1}, \beta_{A_2}, \ldots, \beta_{A_k}, \beta_{A_1 1}, \beta_{A_1 2}, \ldots, \beta_{A_1 l_1}, \ldots, \beta_{A_k 1}, \ldots, \beta_{A_k l_k}). \tag{2.10}$$

The likelihood for estimating the model parameters based on a random sample n individuals as in Eq. (2.1) is given as:

$$L(\boldsymbol{\beta}, \mathbf{Y}) = \prod_{s=1}^{n} \prod_{i=1}^{G} \prod_{j \neq j'} P_{ijj'}^{Y_{isj}}. \tag{2.11}$$

Estimation of the parameters is done in SAS® maximizing the likelihood given in Eq. (2.11).

Attribute and attribute-level data in the experiments are a series of $1's$ and $0's$, indicating the attributes and attribute-levels in the choice pair with positive and negative signs for best and worst attribute-levels, respectively. When fitting a conditional logit model to the data, parameter estimates for the last attribute and last attribute-level for each attribute are not retrievable due to singularity issues. According to Flynn et al. (2007), these parameter estimates are needed to determine the impact of attribute, which is the essential purpose for experiments of this design. To estimate these parameters, the following identifiability condition defined on the parameters of the attribute-levels must be met,

$$\Sigma_{i=1}^{l_k} \beta_i = 0 \tag{2.12}$$

or

$$\beta_{l_k} = -\sum_{j=1}^{l_i - 1} \beta_j \tag{2.13}$$

for all $k = 1, 2, \ldots, K$ (Street and Burgess 2007; Flynn et al. 2007; Grasshoff et al. 2003).

Next, the goal will be to build a functional form of the attributes and the attribute-levels and estimate the associated model parameters that reflect their utilities.

3 Functional Form of Attribute-Level Best-Worst Discrete Choice Model

The attribute-level best-worst DCEs are modified traditional DCEs. Models and theory done for traditional DCEs have not been completely evaluated in terms of best-worst scaling experiments. It is of interest to us to extend the model built on a function of the data as presented in Grasshoff et al. (2003, 2004), and Grossmann et al. (2009) to the attribute-level best-worst DCEs. On extending this work to these experiments, we provide an additional way to define the systematic component that provides flexibility not seen in traditional methods.

Considering functions of the attributes as they enter into the utility function is not a new idea. Van Der Pol et al. (2014) present the systematic components of the utility defined as linear functions, quadratic functions, or as stepwise functions of the attributes. Grasshoff et al. (2013) define the functions as regression functions of the attributes and attribute-levels in the model.

In the attribute-level best-worst DCEs, a set of G profiles, or products, are examined. The profiles are given as $\mathbf{x}_i = (x_{i1}, x_{i2}, \ldots, x_{iK})$, where x_{ij} is the attribute-level in profile $i = 1, 2, \ldots, G$ that corresponds to the jth attribute, where $j = 1, 2, \ldots, K$. The choice task for respondents is to choose the best-worst pair of attribute-levels. In the experiment, respondents make paired comparisons within the profiles instead of between as in traditional DCEs.

In the attribute-level best-worst DCEs, the utility of the pairs is composed of the utility corresponding to the best attribute-level and the worst attribute-level. The regression functions presented in Grasshoff et al. (2003) are applied to the attributes and attribute-levels within the respective systematic components. Let $\mathbf{f}$ be the set of regression functions for the best attribute-levels in the pairs, and $\mathbf{g}$ the set of regression functions for the worst attribute-levels in the pairs. The $p \times 1$ parameter vector $\boldsymbol{\beta}$ still must satisfy the identifiability condition given in Eq. (2.13).

As noted in Marley and Louviere (2005), the inverse random utility model must be used so that the properties are met for the conditional logit model. Taking the systematic component defined in Eq. (2.2), the functional systematic component for the pair $(x_{ij}, x_{ij'})$ is defined as:

$$V_{ijj'} = V_{ij} - V_{ij'} = (\mathbf{f}(\mathbf{x}_{ij'}) - \mathbf{g}(\mathbf{x}_{ij'}))'\boldsymbol{\beta}, \tag{3.1}$$

where $j, j' = 1, 2, \ldots, K, \quad j \neq j'$, and $i = 1, 2, \ldots, G$.

The probability an alternative is chosen depends on the definition of the utility and the distribution of the error terms. Referring back to Eq. (2.5) under the conditional logit, the probability is

$$
\begin{aligned}
BW_{x_i}(x_{ij}, x_{ij'}) &= \frac{\exp(V_{ijj'})}{\displaystyle\sum_{(x_{iq}, x_{iq'}) \in C_i} \exp(V_{iqq'})} \\[2ex]
&= \frac{\exp(V_{ij} - V_{ij'})}{\displaystyle\sum_{(x_{iq}, x_{iq'}) \in C_i} \exp(V_{iq} - V_{iq'})} \\[2ex]
&= \frac{\exp((\mathbf{f}(\mathbf{x}_{ij}) - \mathbf{g}(\mathbf{x}_{ij'}))'\boldsymbol{\beta})}{\displaystyle\sum_{(x_{iq}, x_{iq'}) \in C_i} \exp((\mathbf{f}(\mathbf{x}_{iq'}) - \mathbf{g}(\mathbf{x}_{iq'}))'\boldsymbol{\beta})},
\end{aligned}
\tag{3.2}
$$

where $i = 1, 2, \ldots, G, \quad j, j' = 1, 2, \ldots, K$, and $j \neq j'$.

The forms of the systematic components of the utilities as well as their associated probabilities depend on the definition of the regression functions $\mathbf{f}$ and $\mathbf{g}$. We define the regression functions used in the traditional attribute-level best-worst model and extend the definition of the regression functions to a more general form that provides flexibility in the model. We present that feasible version in the next subsection followed by simulated example.

3.1 Regression Functions Definitions

As presented earlier, best-worst DCEs take into account we provided the design, probabilities, and properties associated with attribute-level best-worst pairs. The data in these experiments are defined as a series of $1's$, $0's$, and $-1's$ corresponding to the best and worst attributes and attribute-levels in a given choice pair. There exist a set of functions $\mathbf{f}$ and $\mathbf{g}$ defined on the attribute-level pair that produces traditional methods.

In the attribute-level best-worst DCEs, a set of G profiles, or products, are examined. The profiles are given as $\mathbf{x}_i = (x_{i1}, x_{i2}, \ldots, x_{iK})$, where x_{ij} is the attribute-level in profile $i = 1, 2, \ldots, G$ that corresponds to the jth attribute for $j = 1, 2, \ldots, K$. Let us consider the choice is given as $(x_{ij}, x_{ij'})$, where $j \neq j'$, $j, j' = 1, 2, \ldots, K$, and $i = 1, 2, \ldots, G$. Let $\mathbf{f}$ be the set of regression functions defined on the best attribute-level in a pair and $\mathbf{g}$ be the set of regression functions defined on the worst attribute-level in a pair.

In the traditional attribute-level best-worst DCE, the regression functions $\mathbf{f}$ and $\mathbf{g}$ are defined as indicator functions. The indicator functions are $p \times 1$ vectors. For the attributes, they are defined as,

$$I_{A_k}(x_{ij}) = \begin{cases} 1, & \text{if } x_{ij} \in A_k, \\ 0, & \text{otherwise}, \end{cases} \tag{3.3}$$

and for the attribute-levels,

$$I_{A_k x_k}(x_{ij}) = \begin{cases} 1, & \text{if } x_{ij} = x_k \text{ for } x_k \in A_k, \\ 0, & \text{otherwise}. \end{cases} \tag{3.4}$$

where $j, k = 1, 2, \ldots, K$ and $i = 1, 2, \ldots, G$.

The best-worst systematic component for the pair $(x_{ij}, x_{ij'})$ is given as,

$$\begin{aligned} V_{ijj'} &= V_{ij} - V_{ij'} \\ &= (\mathbf{f}(\mathbf{x}_{ij}) - \mathbf{g}(\mathbf{x}_{ij'}))'\boldsymbol{\beta} \\ &= \sum_{k=1}^{K} \left[I_{A_k}(x_{ij})\beta_{A_k} + \sum_{j=1}^{l_k} I_{A_k x_{kj}}(x_{ij})\beta_{A_k x_{kj}} \right] \end{aligned}$$

$$-\sum_{k=1}^{K}\left[I_{A_k}(x_{ij'})\beta_{A_k} + \sum_{j'=1}^{l_k} I_{A_k x_{kj'}}(x_{ij'})\beta_{A_k x_{kj'}}\right]$$

$$=\sum_{k=1}^{K}\left[I_{A_k}(x_{ij})\beta_{A_k} - I_{A_k}(x_{ij'})\beta_{A_k}\right.$$

$$\left.+\sum_{j=1}^{l_k} I_{A_k x_{kj}}(x_{ij})\beta_{A_k x_{kj}} - \sum_{j'=1}^{l_k} I_{A_k x_{kj}}(x_{ij'})\beta_{A_k x_{kj'}}\right]$$

$$= I_{A_j}\beta_{A_j} - I_{A_{j'}}\beta_{A_{j'}} + I_{A_j x_{ij}}\beta_{A_j x_{ij}} - I_{A_{j'} x_{ij'}}\beta_{A_{j'} x_{ij'}}, \tag{3.5}$$

where $j, j' = 1, 2, \ldots, K$, $j \neq j'$, and $i = 1, 2, \ldots, G$.

Rewriting the indicator functions of the A_k and $A_k x_k$, a more general form of the regression functions can be defined. Let b_{A_k} and $b_{A_k x_k}$ be constants corresponding to the best attribute and attribute-levels in a pair, and w_{A_k} and $w_{A_k x_k}$ be constants corresponding to the worst attribute and attribute-levels in a pair, where $x_k = 1, 2, \ldots, l_k$ and $k = 1, 2, \ldots, K$. The regression functions $\mathbf{f}$ and $\mathbf{g}$ are given as,

$$\mathbf{f}(x_{ij}) = \sum_{k=1}^{K}\left[b_{A_k} I_{A_k}(x_{ij}) + \sum_{j=1}^{l_k} b_{A_k x_k} I_{A_k x_k}(x_{ij})\right] \tag{3.6}$$

and

$$\mathbf{g}(x_{ij'}) = -\sum_{k=1}^{K}\left[w_{A_k} I_{A_k}(x_{ij'}) + \sum_{j=1}^{l_k} w_{A_k x_k} I_{A_k x_k}(x_{ij'})\right] \tag{3.7}$$

where $j, j' = 1, 2, \ldots, K$, $j \neq j'$, and $i = 1, 2, \ldots, G$.

The above functions are simple linear process which can be used to model the attribute-level best-worst DCEs. Furthermore, the dependence, or functional dependence, can be extended by considering

$$\mathbf{f}(x_{ij}) = \sum_{k=1}^{K}\left[f_k(A_k) + \sum_{j=1}^{l_k} f_{k,j}(A_k x_{kj})\right]. \tag{3.8}$$

and

$$\mathbf{g}(x_{ij}) = -\sum_{k=1}^{K}\left[g_k(A_k) + \sum_{j=1}^{l_k} g_{k,j}(A_k x_{kj})\right]. \tag{3.9}$$

where f_k, g_k, $f_{k,j}$, and $g_{k,j}$ can be linear, nonlinear, or kernel based functional form for best and worst attributes and attributes-levels, respectively.

The regression functions defined in this way provide flexibility than the traditional attribute-level best-worst DCEs. Consumer preference in products is constantly changing, new information about the product comes to light or as trends come and go. Hence, the data collected on a product may be dynamic. The addition of these constants to the regression functions provides researchers the ability to scale the data to reflect current trends or changes in the products. For example, let us consider the products being modeled are pharmaceuticals such as in the contraceptives as proposed in Knox et al. (2012, 2013). If new information about a brand of contraceptives posing a health risk was discovered, then using regression functions, it is possible to update the model to reflect this change. Assuming the change is to remove the brand. The attribute-level associated with the brand may have $b_{kx_k} = w_{kx_k} = 0$, where $x_k = 1, 2, \ldots, l_k$ and $k = 1, 2, \ldots, K$ to represent its removal from the market. For all the pairs this attribute-level was in, the information the choice pair provides in terms of the other attributes and attribute-levels would remain intact. The model would be estimated again and the parameter vector, $\boldsymbol{\beta}$, would provide the updated impact of the attributes and attribute-levels in the experiment.

3.2 Data Example

In the simulated example an empirical setup is considered. We assume $K = 3$ attributes with $l_1 = 2, l_2 = 3$, and $l_3 = 4$ attribute-levels in an unbalanced design. There are $2 \times 3 \times 4 = 24$ possible profiles, or products, in this experiment. The total number of attribute-levels is $L = \sum_{i=1}^{k} l_i = 9$, and the total number of choice pairs is $J = \sum_{k=1}^{K} l_k(L - l_k) = 52$.

We simulated data for $n = 300$ respondents for 24 profiles. Each choice set has $\tau = K(K - 1) = 6$ alternatives to choose from. Using the parameters given in Table 1, we simulated data in R. The data was then exported from R into the SAS® environment. Using the SAS® procedure called MDC (multinomial discrete choice), the conditional logit model was fitted to the data. The parameter estimates for the generated data are given in Table 1. The parameter estimates are close to the original parameters for this example.

We consider an example where the model is built on the regression functions **f** and **g** of the data. We define **f** and **g** as given in Eqs. (3.6) and (3.7). The weights used in the regression functions are given as: $b_{A_1} = w_{A_1} = -2, b_{A_2} = w_{A_2} = 5,$ $b_{A_3} = w_{A_3} = 1, b_{A_11} = w_{A_11} = b_{A_12} = w_{A_12} = -2, b_{A_21} = w_{A_21} = b_{A_22} = w_{A_22} = b_{A_23} = w_{A_23} = 5$ and $b_{A_31} = w_{A_31} = b_{A_32} = w_{A_32} = b_{A_33} = w_{A_33} = b_{A_34} = w_{A_34} = 1.$

The conditional logit model is fit to the data and the resulting parameter estimates are given in Table 1. The parameter estimates provide the adjusted attribute and

Table 1 Parameters and parameter estimates for simulated example

Parameters	β	Estimates		Functional form	
		$\hat{\beta}$	SE	$\hat{\beta}$	SE
β_{A_1}	-2.0000	-2.0711	0.0621	0.9787	0.0289
β_{A_2}	1.5000	1.5248	0.0438	0.3042	0.0082
β_{A_3}	*	*	*	*	*
$\beta_{A_1}1$	-2.0000	-2.0308	0.0619	0.9838	0.0288
$\beta_{A_1}2$	2.0000	2.0308	*	-0.9838	*
$\beta_{A_2}1$	1.9900	2.0970	0.0804	0.3864	0.0148
$\beta_{A_2}2$	-0.2900	-0.3567	0.0482	-0.0548	0.0092
$\beta_{A_2}3$	-1.7000	-1.7403	*	-0.3316	*
$\beta_{A_3}1$	-0.9200	-0.8914	0.0407	-0.8867	0.0410
$\beta_{A_3}2$	-0.1800	-0.1805	0.0368	-0.1806	0.0368
$\beta_{A_3}3$	0.5000	0.4911	0.0369	0.4966	0.0366
$\beta_{A_3}4$	0.6000	0.5808	*	0.5707	*

attribute-level impacts. In Sect. 5, we will evaluate the changes in expected utility for this weighted data in comparison to the original data and model.

We can see the impact of weighting as a reciprocal change in the impact of attribute 1 is noticed as its value goes from -2 to 0.9787. We will use these functional forms and included time in them in Sect. 5 in the simulated example with two scenarios.

We utilize the new definition of the systematic components in the modeling of attribute-level best-worst DCEs across time. In Sect. 4, we extend the work done here to Markov decision processes. The generalized form of the systematic components we provided allows for the evaluation of hypothetical future scenarios.

4 Markov Decision Process

Markov decision processes (MDPs) are sequential decision-making processes. A decision process is said to be Markovian if the future depends on the present and not the past. In that sense, a Markov process is a memoryless practice. MDPs seek to determine the policy, or set of decision rules, under which maximum reward over time is obtained. According to Puterman (2014), decision processes are defined by the set (S, R, D), where S is the set of states, R is the set of rewards, and D is the set of possible decisions for each time step. Let $s_t \in S$ be the states occupied at time t, $r_t(s_t)$ be the rewards associated with s_t, and $d_t(r_t, s_t)$ is the decision based on possible rewards and states at time t, where $0 \leq t \leq T$. The rewards are defined as the expected gain, or loss, associated with the state. With regard to DCEs, the states are the choice pairs and the rewards are the utility associated with the choice in alternative.

The definition of time is important in the methods for mapping the decision processes. These processes may be discrete or continuous in time with finite or infinite horizon. For the purpose of this chapter, our interest is in discrete time, finite horizon MDPs, that is $t = 1, 2, \ldots, T$ where $T < \infty$. Numerical methods such as dynamic programming are used to estimate the expected rewards for this type of MDPs.

As the decision process is Markovian, the transition probability to the next state s_{t+1} based solely on the decision made at the current state, s_t, is $p(s_{t+1}|s_t)$, where $t = 1, 2, \ldots, T$ (Puterman 2014). The transition probabilities are the drivers of this sequential decision-making process. The decision process maps the movement from one state to another over time, t, based on rewards received and the optimal decision set. The optimal decision rule is known as the policy, $\delta = (d_1^*, d_2^*, \ldots, d_T^*)$, where d_t^* is the decision at time $t = 1, 2, \ldots, T$ that yields the maximum expected reward (Puterman 2014).

While there exists some literature on the application of MDPs in traditional DCEs, we have not encountered any work in the literature to extend these methods to best-worst scaling DCEs. In this chapter, we extend the use of MDPs to Case 2 of best-worst scaling models, the attribute-level best-worst DCEs.

In traditional MDPs, the value functions are computed for each of the J alternatives, or products. At each time point, $t = 1, 2, \ldots, T$, the decision d_t is to choose the alternative that provides the maximum expected utility given information about the state $s_t = (\mathbf{x}_t, \epsilon_t)$, where $\mathbf{x}_t$ is the set of K attributes. The decision made is between alternatives in the traditional DCEs. In attribute-level best-worst DCEs, the experiments model choices within products not between products.

In attribute-level best-worst DCEs, there are K attributes describing a product each with l_k levels, where $k = 1, 2, \ldots, K$. The total number of products in these experiments is $\prod_{k=1}^{K} l_k$. The products are represented in the experiment by a profile. The profile corresponding to the ith product is given as $\mathbf{x}_i = (x_{i1}, x_{i2}, \ldots, x_{iK})$, where x_{ik} is the attribute-level corresponding to the attribute A_k for $k = 1, 2, \ldots, K$ for $i = 1, 2, \ldots, G$. Within each choice set there are $\tau = K(K - 1)$ choices. A respondent is asked to evaluate G choice sets in the experiment.

MDPs model the decision process for respondents over multiple time points. For attribute-level best-worst DCEs, the model is built within the choice sets corresponding to each of the G choice sets. In traditional DCEs, there are J alternatives evaluated at each time point producing J value functions at each time point. Attribute-level best-worst DCEs require a respondents to evaluate a series of G choice sets each with τ choices, thus there are τ value functions for each choice set in attribute-level best-worst MDPs. Our interest is to further model the sequence of decisions made by introducing the time element into the experiments. For attribute-level best-worst DCEs, we consider discrete time finite horizon MDPs where:

- G choice sets are modeled across time.
- $\mathbf{x}_{ijj'}^t = (x_{ij}, x_{ij'})$ are the attributes and attribute-levels corresponding to the choices in set $\mathcal{C}_i$, $i = 1, 2, \ldots, G$, $j \neq j'$, $j, j' = 1, 2, \ldots, K$, and $t = 1, 2, \ldots, T$.

- The decision set depends on the choice set, called D_i, and we evaluate $d_i^t \in D_i$, where $i = 1, 2, \ldots, G$ and $1 \leq d_{ti} \leq \tau$.
- The set of possible states in the experiment depends on the choice set, called S_i, where $s_i^t = (x_{ij}^t, x_{ij'}^t) \in S_i$, where $j \neq j'$, $j, j' = 1, 2, \ldots, \tau$, $i = 1, 2, \ldots, G$, and $1 \leq s_{ti} \leq \tau$.
- Transition probabilities depend on a set of parameters $\boldsymbol{\theta}$ that are assumed known, or estimated from data (Arcidiacono and Ellickson 2011).
- Transition probability matrices, $P^t_{s_i s_i'}$, are dependent on the choice set being evaluated.

In attribute-level best-worst DCEs, the MDPs model the choices in attribute-level pairs within choice sets over time. Therefore, the transition probabilities and value functions must be defined within the choice sets. Bellman (1954) utilized dynamic programming to evaluate the value function, also known as Bellman's equation, at each time step. Rust (1994, 2008) presented the use of dynamic programming for evaluating DCEs as MDPs.

The value function for DCEs defined by Bellman's equation is given as:

$$V_t(x_t, \epsilon_t) = \max_{d_t \in D} \sum_{t'=t}^{T} P^{t'}_{ss'} \left[\gamma^{t'-t} U(\mathbf{x}_{t'}, d_{t'}) + \epsilon(d_{t'}) | \mathbf{x}_t \right]$$

$$= \max_{d_t \in D} E \left(\sum_{t'=t}^{T} \gamma^{t'-t} U(\mathbf{x}_{t'}, d_{t'}) + \epsilon(d_{t'}) | \mathbf{x}_t, \epsilon_t \right), \tag{4.1}$$

where $d_t \in D$ is the decision at time t, $U(\mathbf{x}_t, d_t)$ is the derived iterated/expected utility, ϵ_t is the associated error term at time t, where $t = 1, 2, \ldots, T$, and discount utility rate is given by $\gamma \in (0, 1)$ (Bellman 1954).

The value functions are computed recursively via dynamic programming. To determine the value function, backwards recursion must be used. At the last time point, T, the value function is the utilities associated with the different states. Adopting for the attribute-level best-worst DCEs eventually the value function is given by,

$$V_i^t(s_i^t, d_i^t) = U(s_i^t, d_i^t) + \sum_{s_i'^{t+1} \in S_i} \gamma V_{t+1}^i(s_i'^{t+1}, d_i^{t+1}) P^t_{s_i s_i'}, \tag{4.2}$$

where $t = 1, 2, \ldots, T$, $U(s_i^t, d_i^t)$ represents the utility associated with the state s_i^t and decision d_i^t, and discount utility rate is given by $\gamma \in (0, 1)$ and $i = 1, 2, \ldots, G$. The decision $d_i^t = (x_{ij}, x_{ij'})$ is a choice pair within C_i, where $i = 1, 2, \ldots, G$, $j, j' = 1, 2, \ldots, K$, and $j \neq j'$. In the attribute-level best-worst DCEs, there will be $\tau = K(K-1)$ value functions per each of the G choice sets. One of the disadvantages of these experiments is the "curse of dimensionality" (Rust 2008). As the number of attributes, attribute-levels, and profiles grow in the experiment, the estimation process becomes exponentially more difficult as dynamic programming requires an explicit discretization of the states, decisions, and the value function as seen in Eq. (4.2) depends on the utility and transition probabilities over time. The ability to direct the system, via the transition probabilities, when it is of a higher dimension becomes difficult, if not impossible.

In the following subsections, we provide definitions and insights regarding these components to the value function. In Sect. 3, we provided a functional form of the utility that we can apply in these time dependent processes. Furthermore, we define dynamic transition probabilities that we apply to attribute-level best-worst DCEs. In Sect. 5, simulations of MDPs for the attribute-level best-worst DCEs are provided.

4.1 Utility

Marschak (1960) presented random utility theory defining utility to include a systematic component V_{ij} and an unobserved component ϵ_{ij} for $i = 1, 2, \ldots, n$ and $j = 1, 2, \ldots, J$. A consumer chooses the alternative that provides them with the maximum utility. The utility function for traditional DCEs is given as:

$$U_{ij} = V_{ij} + \epsilon_{ij} = \mathbf{x}'_{ij}\boldsymbol{\beta} + \epsilon_{ij},$$

for $i = 1, 2, \ldots, n$ and $j = 1, 2, \ldots, J$. For MDPs in the traditional DCEs, the utility is then,

$$U_t(\mathbf{x}_t, d_t) = \mathbf{x}'_t\boldsymbol{\beta} + \epsilon_t, \tag{4.3}$$

where $t = 1, 2, \ldots, T$. Common models to determine the parameter estimates of $\boldsymbol{\beta}$ are conditional logit, generalized extreme value distributions, and probit models. We consider the conditional logit model in this chapter. Stenberg et al. (2007) provided that the definition of utility/reward in MDPs maybe constant over time or time dependent/dynamic in nature.

The definition of utility in attribute-level best-worst DCEs that meets the necessary independence from irrelative alternative (IIA) condition (Luce 1959) for the conditional logit model is given in Eq. (2.2) under the inverse random utility theory presented in Marley and Louviere (2005). For the corresponding choice pair $x_{ijj'}^t = (x_{ij}^t, x_{ij'}^t) \in C_i^t$ the corresponding utility is given as,

$$U_{ijj'}^t = V_{ij}^t - V_{ij'}^t + \epsilon_{ijj'}^t = V_{ijj'}^t + \epsilon_{ijj'}^t,$$

where $i = 1, 2, \ldots, G, j, j' = 1, 2, \ldots, K$, and $j \neq j'$.

Referring back to Sect. 3, the systematic component is defined as a model built on functions of the best and worst attribute-levels in the pair, using Eq. (3.1),

$$V_{ijj'} = (\mathbf{f}_t(\mathbf{x}_{ij}) - \mathbf{g}_t(\mathbf{x}_{ij'}))'\boldsymbol{\beta}, \tag{4.4}$$

where $\boldsymbol{\beta}$ are the attribute and attribute-level coefficients, $j, j' = 1, 2, \ldots, K$, $j \neq j'$, $i = 1, 2, \ldots, G$, and $\mathbf{f}_t$ and $\mathbf{g}_t$, $t = 1, 2, \ldots, T$, are regression functions defined on the best and worst attributes and attribute-levels, respectively. In the traditional attribute-level best-worst DCEs, the functions $\mathbf{f}_t$ and $\mathbf{g}_t$ are given as indicator functions of the best and worst attributes and attribute-levels as shown in Eq. (3.5).

However, using this functional form of the systematic component, we may consider alternative definitions of the systematic component. In Sect. 3, we provide a weighted function for $\mathbf{f}_t$ and $\mathbf{g}_t$, given in Eqs. (3.6) and (3.7). Let $b^t_{A_k}$ and $b^t_{A_k x_k}$ be weights corresponding to the best attribute and attribute-levels in a pair, and $w^t_{A_k}$ and $w^t_{A_k x_k}$ be weights corresponding to the worst attribute and attribute-levels in a pair, where $x_k = 1, 2, \ldots, l_k$, $k = 1, 2, \ldots, K$, and $t = 1, 2, \ldots, T$. The regression functions $\mathbf{f}_t$ and $\mathbf{g}_t$ are given as,

$$\mathbf{f}_t(x_{ij}) = \sum_{k=1}^{K} \left[b^t_{A_k} I_{A_k}(x_{ij}) + \sum_{j=1}^{l_k} b^t_{A_k x_k} I_{A_k x_k}(x_{ij}) \right], \qquad (4.5)$$

and

$$\mathbf{g}_t(x_{ij'}) = -\sum_{k=1}^{K} \left[w^t_{A_k} I_{A_k}(x_{ij'}) + \sum_{j=1}^{l_k} w^t_{A_k x_k} I_{A_k x_k}(x_{ij'}) \right], \qquad (4.6)$$

where $j, j' = 1, 2, \ldots, K$, $j \neq j'$, and $i = 1, 2, \ldots, G$.

Defining the systematic components according to the weighted function allows the utility to change over time. We considered in Sect. 3 an example where an attribute-level no longer exists in the future. The weighted functions of $\mathbf{f}_t$ and $\mathbf{g}_t$ allowed us to update the parameter estimates, thus the utilities, using these weights. It is conceivable in the future that an attribute-level scale may need to be adjusted for possible bettering, worsening, or removal type of conditions for that attribute-level.

4.2 Transition Probabilities

MDPs have infinitely many possible futures able to be considered in the simulations. The definition of transition probabilities is the vehicle that drives the processes to these different futures. However, determining transition probabilities for MDPs is a difficult task. One way for estimating the transition probabilities is using maximum likelihood estimates (MLEs). An empirical solution to the transition probabilities may be determined by considering the transition probabilities as a multinomial distribution (Lee et al. 1968).

In the attribute-level best-worst DCEs, there are τ choices within a choice set $\mathcal{C}_i$, where $i = 1, 2, \ldots, G$. There are τ states, and/or decisions, possible at each of the time points. The transition probabilities are denoted as $P_{ss'} = P(s_{t+1} = s' | s_t = s)$, where $s_t, s_{t+1} \in S$ and $S = \{1, 2, \ldots, \tau\}$. Let N_i be the respondents common to time t and $t + 1$ in the experiment and $n_{iss'}$ be the number of respondents who chose s at time t and s' at $t + 1$, where $t = 1, 2, \ldots, T$ and $i = 1, 2, \ldots, G$. The transition choice probability is given by the multinomial distribution as:

$$f(p_{is1}, p_{is2}, \ldots, p_{is\tau}) = \frac{N_i!}{n_{is1}! n_{is2}! \ldots n_{is\tau}!} p_{is1}^{n_{is1}} p_{is2}^{n_{is2}} \cdots p_{is\tau}^{n_{is\tau}}, \qquad (4.7)$$

and log-likelihood is given as

$$log(L) = log\left(\frac{N_i!}{\prod\limits_{s'=1}^{\tau} n_{iss'}!}\right) + \sum_{s'=1}^{\tau} n_{iss'} log(p_{iss'}),$$

where $s = 1, 2, \ldots, \tau$, $i = 1, 2, \ldots, G$, $p_{iss'} \geq 0$, and $\sum\limits_{s'=1}^{\tau} p_{iss'} = 1$.

Due to the constraint $\sum\limits_{s'=1}^{\tau} p_{iss'} = 1$, Lagrange multipliers, λ, are used and the Lagrangian function is given as:

$$G(p_{ss'}) = LL(p_{iss'}) - \lambda\left(\sum_{s'=1}^{\tau} p_{iss'} - 1\right),$$

where $s = 1, 2, \ldots, \tau$, $i = 1, 2, \ldots, G$, $p_{iss'} \geq 0$, and $\sum\limits_{s'=1}^{\tau} p_{iss'} = 1$.

Taking the partial derivative of the Lagrangian to determine the MLEs gives us $\frac{n_{iss'}}{\lambda} = p_{iss'}$ where $s' = 1, 2, \ldots, \tau$. Under the constraint, $\sum\limits_{s'=1}^{\tau} p_{iss'} = 1$, the value of $\lambda = \sum\limits_{s'=1}^{\tau} n_{iss'} = N_i$. Thus, the MLE for $p_{iss'} = \frac{n_{iss'}}{N_i}$ for $s, s' = 1, 2, \ldots, \tau$ and $i = 1, 2, \ldots, G$.

The MLE of $p_{iss'}$ is computationally simple; however, access to the information needed to compute it may not always be available. To compute the MLE of this nature, we would need to have respondents evaluate the same choice sets at two time periods, which is not necessarily an easy task. Furthermore, this is considering the transition matrix is stationary. It is possible to consider a dynamic transition matrix that changes over time, that is $p_{iss'}^t$ for $t = 1, 2, \ldots, T$. A transition matrix of this nature would need to have multiple time periods of data for the same respondents evaluating the same choice sets to compute the empirical probabilities. Instances where multiple time periods of data for respondents are not possible, one must consider alternative methods for determining the transition probabilities.

There are infinitely many possible transition probabilities in MDPs. Common methods for determining these probabilities is to take a Bayesian approach and the other is a rational observation according to Rust (2008). Arcidiacono and Ellickson (2011) indicates that the transition probabilities, $P_{ss'} = P(s_{t+1} = s'|s_t = s, \boldsymbol{\theta})$ are a probability function, where the parameters $\boldsymbol{\theta}$ are assumed known. Rust (1994, 2008) state that discrete decision processes, as we are considering in the attribute-level best-worst models, the transition parameters and probabilities are often times non-parametrically identified. Chadès et al. (2014) applied MDPs to solve problems in an ecological setting. As they mentioned, to suggest guidance in transition probabilities would require running several scenarios. To the best of our knowledge, such technique has not yet been applied to consumer choice experiments with attribute and attribute-level best-worst experiments.

We provide a definition of the parameters for the transition probabilities under the rational observation that may be used in stationary or dynamic transition matrices. This method maintains the researcher's ability to guide the MDPs in the direction of their choosing where the transitions occur at a rate determined by the researcher. In such a way, the researcher is able to consider stationary or dynamic transition probabilities to model an evolving MDP over time. The researcher may also determine the amount of time points necessary for the system to converge to the decision they were working towards.

In attribute-level best-worst choice models, a set of G choice sets are considered in the experiment. In MDPs, there exists a set of states $s_t \in S$ and possible decisions in $d_t \in D$ for $t = 1, 2, \ldots, T$. For attribute-level best-worst MDPs, the possible states in each choice set are the alternatives, and the decision made at each time point will also be one of the alternatives. For choice set C_i the state s_{ti} and decision d_{ti} are such that $1 \leq s_{ti}, d_{ti} \leq \tau$ where $i = 1, 2, \ldots, G$ and $t = 1, 2, \ldots, T$.

Each state $s = s_{t_i}$ gives rise to a new set of states s' at time $t + 1$ with transition probability denoted as $P^t_{iss'} = P^t(s'|s, \boldsymbol{\theta}^t_s)$, where

$$\boldsymbol{\theta}^t_s = (\theta^t_{sA_1}, \theta^t_{sA_2}, \ldots, \theta^t_{sA_K}, \theta^t_{sA_11}, \ldots, \theta^t_{sA_Kl_k})$$

where sA_i indicates involvement of attribute i, and sA_il_k indicates involvement of attribute-level k in attribute i where $i = 1, 2, \ldots, K$ and $k = 1, 2, \ldots, l_k$.

The parameter estimates determined by fitting the conditional logit model, as described in Sect. 3, produced $\hat{\boldsymbol{\beta}}$ a $p = K + \sum_{k=1}^{K} l_k$ vector. These parameter estimates measure the relative impact of each attribute and attribute-level in the decisions made by respondents. The parameters $\boldsymbol{\theta}^t_s$ are the assumed impacts of the attributes and attribute-levels in respondents decisions given they currently occupy state s. We define these parameters as functions of the parameter estimates $\hat{\boldsymbol{\beta}}$, where there is a rate of change in the impacts over time. We define

$$\hat{\boldsymbol{\theta}}^t_s = (a_{sA_1}(t)\hat{\beta}_{A_1}, \ldots, a_{sA_K}(t)\hat{\beta}_{A_k},$$
$$a_{sA_11}(t)\hat{\beta}_{A_11}, \ldots, a_{sA_Kl_k}(t)\hat{\beta}_{A_Kl_k}),$$

where $a's$ are the time factor change and $\beta's$ are fixed for $i = 1, 2, \ldots, G, 1 \leq s \leq \tau$, and $t = 1, 2, \ldots, T$. The definition of the $\mathbf{a}_s(t)$ depends on the state s and time $t = 1, 2, \ldots, T$. We have considered $a_{sj}(t) = a_{sj}^t$, where if $|a_{sj}| < 1$ the impact of the attribute or attribute-level would be lessening with time, where $j = 1, 2, \ldots, K$. If $a_{sj}(t)\hat{\beta}_j = a_{sj}^t \hat{\beta}_j > 0$, then the attribute or attribute-level has a positive impact evolving at the rate a_{sj}^t over time for $j = 1, 2, \ldots, K$ and $t = 1, 2, \ldots, T$. A static, or non-time dependent, system is considered if $a_{sj}(t) = 1$, where $j = 1, 2, \ldots, K$ and $t = 1, 2, \ldots, T$.

As mentioned, these $a_{sj}(t)$ are rates of change that guide the dynamic transition of the decision process. We can easily consider them to be non-time dependent, $a_{sj}(t) = a_{sj}$, defining the transition probabilities as stationary over time. As was mentioned earlier, there are infinitely many possibilities in how we define the transitions. Rust (2008) states that using rational observation to define the transitions of any possible choice behavior on the respondents is possible. Chadès et al. (2014) recommend running many scenarios to determine the transition probabilities that will maximize the expected reward. Our definition also offers infinitely many possibilities in terms of the definition; however, we defined a rate of change to consider an evolving system. In this way, the researcher can determine what they consider feasible rates and see if the system eventually evolves to the decision they desire and how long it would take to get there.

Now the transition probability is given as,

$$P^t(s'_{ijj'}|s_i, \boldsymbol{\theta}^t_{s_i}) = P(U^t_{ijj'} > U^t_{ikk'}, \forall k \neq k' \in C_i|s_i, \boldsymbol{\theta}^t_{s_i})$$

$$= P^t(V^t_{ijj'} + \epsilon^t_{ijj'} > V^t_{ikk'} + \epsilon^t_{ikk'}, \forall k \neq k' \in C_i|s_i, \boldsymbol{\theta}^t_{s_i})$$

$$= P^t(\epsilon^t_{ikk'} < \epsilon^t_{ijj'} + V^t_{ijj'} - V^t_{ikk'}, \forall k \neq k' \in C_i|s_i, \boldsymbol{\theta}^t_{s_i}), \quad (4.8)$$

where $j \neq j'$, $j, j' = 1, 2, \ldots, \tau$, $i = 1, 2, \ldots, G$, and $t = 1, 2, \ldots, T$. If we assume the random error terms are independently and identically distributed as type I extreme value distribution, the probability would then be found using the conditional logit, and is given as:

$$P^t(s'_{ijj'}|s_i, \boldsymbol{\theta}^t_{s_i}) = P^t(U^t_{ijj'} > U^t_{ikk'}, \forall k \neq k' \in C_i|s_i, \boldsymbol{\theta}^t_{s_i})$$

$$= \frac{\exp(V^t_{ijj'})}{\sum_{k,k'\in C_i} \exp(V^t_{ikk'})}$$

$$= \frac{\exp((\mathbf{f}_t(\mathbf{x}_{ij}) - \mathbf{g}_t(\mathbf{x}_{ij'}))'\boldsymbol{\theta}^t_{s_i})}{\sum_{k,k'\in C_i} \exp((\mathbf{f}_t(\mathbf{x}_{ik}) - \mathbf{g}_t(\mathbf{x}_{ik'}))'\boldsymbol{\theta}^t_{s_i})}, \quad (4.9)$$

where $j \neq j', k \neq k', j, j' = 1, 2, \ldots, \tau, i = 1, 2, \ldots, G$, and $t = 1, 2, \ldots, T$.

The transition matrix is then a $\tau \times \tau$ matrix of the form,

$$
P^t = \begin{pmatrix}
P^t_{i11} & P^t_{i12} & \cdots & P^t_{i1\tau} \\
P^t_{i21} & P^t_{i22} & \cdots & P^t_{i2\tau} \\
\cdot & \cdot & \cdots & \cdot \\
\cdot & \cdot & \cdots & \cdot \\
\cdot & \cdot & \cdots & \cdot \\
P^t_{i\tau 1} & P^t_{i\tau 2} & \cdots & P^t_{i\tau\tau}
\end{pmatrix} = \left(P^t_{iss'} \right)
$$

where $i = 1, 2, \ldots, G$, $s, s' = 1, 2, \ldots, \tau$, and where $\sum_{s'=1}^{\tau} P^t_{iss'} = 1$. The transition matrix may be either stationary or dynamic in nature. In our definition of $\theta^t_{s_i}$, this is determined by the rate $a_{s_i j}(t)$, where $i = 1, 2, \ldots, G, 1 \le j \le p$, and $t = 1, 2, \ldots, T$. In Sect. 5, we provide simulations under stationary and dynamic transition probabilities and make comparisons.

5 Simulation Example

In the simulated example, an empirical setup is considered. We assume $K = 3$ attributes with $l_1 = 2, l_2 = 3$, and $l_3 = 4$ attribute-levels in an unbalanced design. There are $2 \times 3 \times 4 = 24$ possible profiles, or products, in this experiment. The total number of attribute-levels is $L = \sum_{i=1}^{k} l_i = 9$, and the total number of choice pairs is $J = \sum_{k=1}^{K} l_k(L - l_k) = 52$.

Louviere and Woodworth (1983), Street and Knox (2012), and Grasshoff et al. (2004) discussed the benefits in using orthogonal arrays. Generally, orthogonal experimental designs are utilized in attribute-level best-worst DCEs due to the large number of profiles in a full factorial design. There is a package in R called DoE.design that creates full factorial and orthogonal designs for a given set of attributes and attribute-levels. To obtain an orthogonal design, the oa.design function is used. For this experiment, the orthogonal design returned the full factorial design, so we used the full set of 24 profiles when simulating this data.

We simulated data for $n = 300$ respondents for 24 profiles. Each choice set has $\tau = K(K - 1) = 6$ choices to choose from. Using the parameters given in Table 1, we simulated data in R. The data was then exported from R into the SAS® environment. Using the SAS® procedure called MDC (multinomial discrete choice), the conditional logit model was fitted to the data. The parameter estimates for the generated data are given in Table 1. The parameter estimates are close to the original parameters for this example. Using the parameter estimates, the choice utilities were computed and are used to determine the expected utility/value function. The best

Table 2 Choice pairs with the highest utility in the experiment

Best attribute	Level	Worst attribute	Level	Utility
2	1	1	1	12.3633
2	2	1	1	8.8012
3	4	1	1	7.6931

Table 3 Choice pairs with the lowest utility in the experiment

Best attribute	Level	Worst attribute	Level	Utility
1	1	2	1	-9.2594
1	1	2	2	-6.5358
1	1	3	4	-5.7929

and worst 3 choice pairs along with their utilities are presented in Tables 2 and 3, respectively. The opposite of the pairs with the highest utilities have the lowest utilities.

5.1 Scenario 1

We ran the simulation under this scenario with an advantageous proposed structure. The intent is to validate/justify our relative performance over time under stationary sparsity.

In this example, respondents are assumed to make similar decisions at each decision epoch that they made at the previous time point. The transition parameters $\theta^t_{s_i}$ where $s^t_i = (x_{ij}, x_{ij'})$ are defined as for the attributes as,

$$\theta^t_{s_i A_k} = \begin{cases} 1.7|\beta_{A_k}|, & \text{if } x_{ij} \in A_k, \\ -1.7|\beta_{A_k}|, & \text{if } x_{ij'} \in A_k, \\ \beta_{A_k}, & \text{otherwise,} \end{cases} \tag{5.1}$$

and for the attribute-levels,

$$\theta^t_{s_i A_k x_{ik}} = \begin{cases} 1.7|\beta_{A_k x_{ik}}|, & \text{if } x_{ij} = x_{ik} \text{ where } x_{ik} \in A_k, \\ -1.7|\beta_{A_k x_{ik}}|, & \text{if } x_{ij'} = x_{ik} \text{ where } x_{ik} \in A_k, \\ \beta_{A_k x_{ik}}, & \text{otherwise,} \end{cases} \tag{5.2}$$

where $j \neq j'$, $j, j', k = 1, 2, \ldots, K$, $1 \leq x_k \leq l_k$, and $i = 1, 2, \ldots, G$. The transition parameters do not change with time, so the transition matrix is stationary. The goal of this scenario was to design the transition probabilities in a way that the choice made at t is most likely to be made at $t + 1$. If we considered $a_{s_i m}(t) = \beta_m$ for $i = 1, 2, \ldots, G$, and $m = 1, 2, \ldots, p$, then the system would remain static and every row of the transition matrix would be the same. Recall that $p = K + \sum_{k=1}^{K} l_k = 12$ is the number of parameters. We consider $1.7|\beta_m|$ when a state or choice pair at time $t + 1$ has the same best attribute and attribute-level as the state occupied at time t, and $-1.7|\beta_m|$ when a state or choice pair at time $t + 1$ has the same worst attribute and attribute-level as the state occupied at time t. We consider $|\beta_m|$ to control the direction of the impact making sure it is positive for the best attribute and attribute-level of s_i and use $-|\beta_m|$ to make sure its negative for the worst attribute and attribute-level of s_i. We use 1.7 to increase the impact of the best and worst attributes and attribute-levels of s_i. The definition of $a_{s_i m}(t)$ in this way insures that states with common best and worst attributes and attribute-levels as the present state occupied, $s_i^t = (x_{ij}, x_{ij'})$, have a greater probability of being transitioned to, where $i = 1, 2, \ldots, G$, $j \neq j'$, $j, j' = 1, 2, \ldots, K$, and $t = 1, 2, \ldots, T$. The weights associated with the attributes and attribute-levels are selected as: $b_{A_1} = w_{A_1} = -2$, $b_{A_2} = w_{A_2} = 5$, $b_{A_3} = w_{A_3} = 1$, $b_{A_1 1} = w_{A_1 1} = b_{A_1 2} = w_{A_1 2} = -2$, $b_{A_2 1} = w_{A_2 1} = b_{A_2 2} = w_{A_2 2} = b_{A_2 3} = w_{A_2 3} = 5$ and $b_{A_3 1} = w_{A_3 1} = b_{A_3 2} = w_{A_3 2} = b_{A_3 3} = w_{A_3 3} = b_{A_3 4} = w_{A_3 4} = 1$.

Referring back to Sect. 4, the systematic component as a function of the best and worst attribute-level in the pair is as in Eq. (4.4),

$$V_{ijj'} = (\mathbf{f}_t(\mathbf{x}_{ij'}) - \mathbf{g}_t(\mathbf{x}_{ij'}))'\boldsymbol{\beta},$$

where $\mathbf{f}_t$ and $\mathbf{g}_t$, as in Eqs. (4.5) and (4.6) with profile choice pairs shown in Fig. 2.

The value function/expected utilities for Profile 1 are displayed in Fig. 3, with legend displayed in Fig. 2 along with the difference in the value functions over time. Choice pair (x_{22}, x_{12}), where x_{22} is the 2nd level of attribute 2 is the best and x_{12} is the 2nd level of attribute 1 is the worst, is the choice with the highest expected utility. The opposite pair (x_{12}, x_{22}) is the worst choice pair. The pair (x_{34}, x_{22}) has a sharp drop between time $t = 3$ and $t = 4$ because of the change in the weights applied to the attributes and attribute-levels from Eq. (3.1).

The model applied here views the attribute-level best-worst DCEs as sequential leading to partial separation best-worst choices over time. Validity is guided by the transition probabilities under Scenario 1, the participants follow the same choice preferences. In Table 4, the transition probabilities are generally highest on the

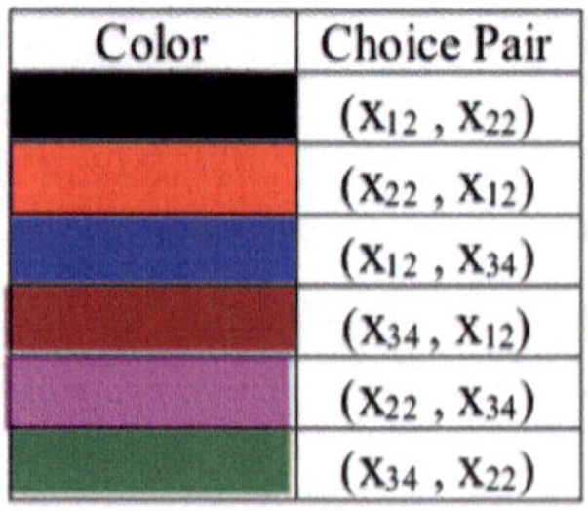

Fig. 2 Legend corresponding to Fig. 3

Color	Choice Pair
	(X_{12}, X_{22})
	(X_{22}, X_{12})
	(X_{12}, X_{34})
	(X_{34}, X_{12})
	(X_{22}, X_{34})
	(X_{34}, X_{22})

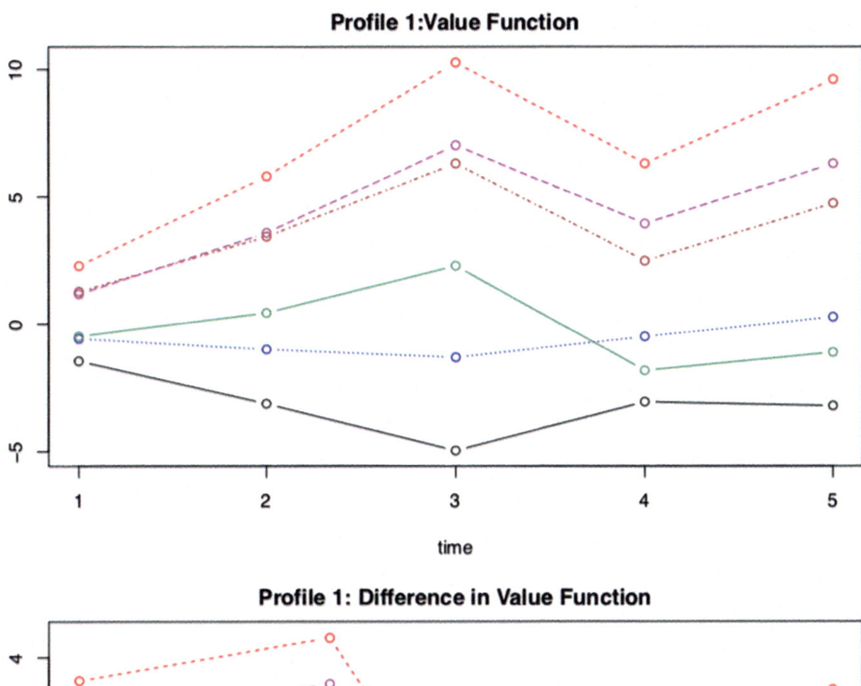

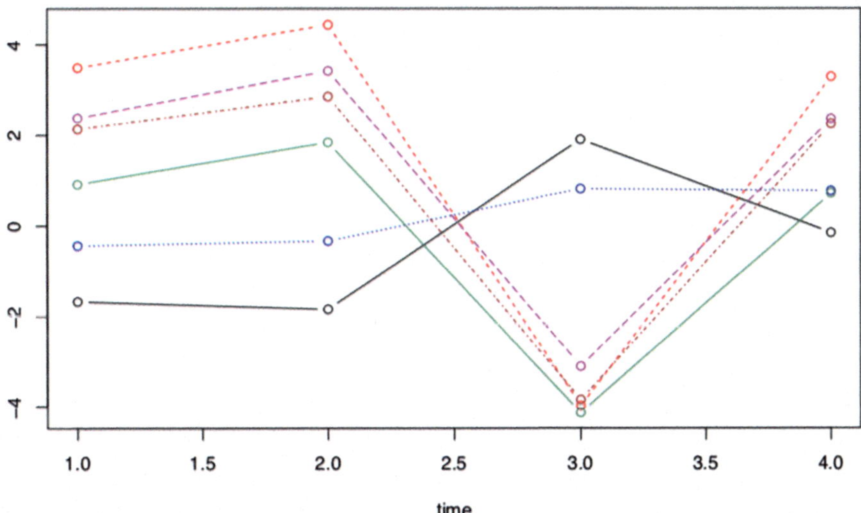

Fig. 3 Expected discounted utility and their differences over time for Profile 1

Table 4 Stationary transition matrix in Scenario 1 for Profile 1

(x_{12}, x_{22})	0.9837	0.0000	0.0136	0.0000	0.0000	0.0027
(x_{22}, x_{12})	0.0000	0.8924	0.0000	0.1074	0.0003	0.0000
(x_{12}, x_{34})	0.0038	0.0000	0.9932	0.0000	0.0030	0.0000
(x_{34}, x_{12})	0.0000	0.0038	0.0000	0.9613	0.0000	0.0003
(x_{22}, x_{34})	0.0000	0.4289	0.0004	0.0002	0.5705	0.0000
(x_{34}, x_{22})	0.0001	0.0004	0.0000	0.7113	0.0000	0.2882

diagonal and the same at each time period as we would expect in this setup. As expected the trend in the utility is kept.

5.2 Scenario 2

In Scenario 2, respondents are allowed to make similar decisions at each time epoch with a different rate of change, making the transition probabilities dynamic. The transition parameters $\theta^t_{s_i}$ where $s^t_i = (x_{ij}, x_{ij'})$ are defined as for the attributes as,

$$
\theta^t_{s_i A_k} = \begin{cases} 1.7^t |\beta_{A_k}|, & \text{if } x_{ij} \in A_k, \\ -1.7^t |\beta_{A_k}|, & \text{if } x_{ij'} \in A_k, \\ \beta_{A_k}, & \text{otherwise}, \end{cases} \tag{5.3}
$$

and for the attribute-levels,

$$
\theta^t_{s_i A_k x_{ik}} = \begin{cases} 1.7^t |\beta_{A_k x_{ik}}|, & \text{if } x_{ij} = x_{ik} \text{ where } x_{ik} \in A_k, \\ -1.7^t |\beta_{A_k x_{ik}}|, & \text{if } x_{ij'} = x_{ik} \text{ where } x_{ik} \in A_k, \\ \beta_{A_k x_{ik}}, & \text{otherwise}, \end{cases} \tag{5.4}
$$

where $j \neq j'$, $j, j', k = 1, 2, \ldots, K$, $1 \leq x_k \leq l_k$, and $i = 1, 2, \ldots, G$.

We ran the simulation under this scenario with advantageous proposed hybrid structure as shown above using the functional form as described in Scenario 1 with profile choice pairs shown in Fig. 4. The transition matrix at time $t = 1$ is kept the same as it was Scenario 1 in Table 4, and subsequent transition probabilities at time $t = 2, 3$, and 4 are given in Tables 5, 6, and 7, respectively. The transition probabilities are highest on the diagonal verifying the direction we wanted in the transitions. The value function/expected utilities for Profile 1 are displayed in Fig. 5, with legend displayed in Fig. 4 along with the difference in value functions. Choice pair (x_{22}, x_{12}), where x_{22} is the 2nd level of attribute 2 is the best and x_{12} is the 2nd

Fig. 4 Legend corresponding to Fig. 5

Color	Choice Pair
	(X_{12}, X_{22})
	(X_{22}, X_{12})
	(X_{12}, X_{34})
	(X_{34}, X_{12})
	(X_{22}, X_{34})
	(X_{34}, X_{22})

Table 5 Dynamic transition matrix in Scenario 2 at time $t = 2$ for Profile 1

(x_{12}, x_{22})	0.9985	0.0000	0.0015	0.0000	0.0000	0.0000
(x_{22}, x_{12})	0.0000	0.9873	0.0000	0.0127	0.0000	0.0000
(x_{12}, x_{34})	0.0002	0.0000	0.9998	0.0000	0.0000	0.0000
(x_{34}, x_{12})	0.0000	0.0019	0.0000	0.9981	0.0000	0.0000
(x_{22}, x_{34})	0.0000	0.0337	0.0001	0.0000	0.9663	0.0000
(x_{34}, x_{22})	0.0000	0.0000	0.0000	0.2082	0.0000	0.7918

Table 6 Dynamic transition matrix in Scenario 2 at time $t = 3$ for Profile 1

(x_{12}, x_{22})	1.0000	0.0000	0.0000	0.0000	0.0000	0.0000
(x_{22}, x_{12})	0.0000	0.9997	0.0000	0.0003	0.0000	0.0000
(x_{12}, x_{34})	0.0000	0.0000	1.0000	0.0000	0.0000	0.0000
(x_{34}, x_{12})	0.0000	0.0000	0.0000	1.0000	0.0000	0.0000
(x_{22}, x_{34})	0.0000	0.0002	0.0000	0.0000	0.9998	0.0000
(x_{34}, x_{22})	0.0000	0.0000	0.0000	0.0058	0.0000	0.9942

Table 7 Dynamic transition matrix in Scenario 2 at time $t = 4$ for Profile 1

(x_{12}, x_{22})	1.0000	0.0000	0.0000	0.0000	0.0000	0.0000
(x_{22}, x_{12})	0.0000	1.0000	0.0000	0.0000	0.0000	0.0000
(x_{12}, x_{34})	0.0000	0.0000	1.0000	0.0000	0.0000	0.0000
(x_{34}, x_{12})	0.0000	0.0000	0.0000	1.0000	0.0000	0.0000
(x_{22}, x_{34})	0.0000	0.0000	0.0000	0.0000	1.0000	0.0000
(x_{34}, x_{22})	0.0000	0.0000	0.0000	0.0000	0.0000	1.0000

level of attribute 1 is the worst, still remains the choice with the highest expected utility as in Scenario 1. The opposite pair (x_{12}, x_{22}) is the worst choice pair. We also notice more shifts in expected utility than in the previous scenarios for Profile 1. Scaling the data makes the utilities shift in much more extreme values.

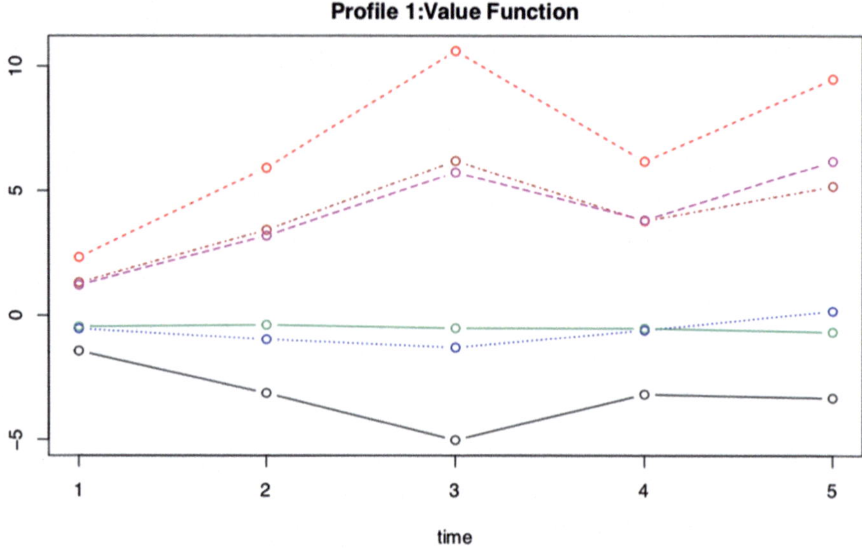

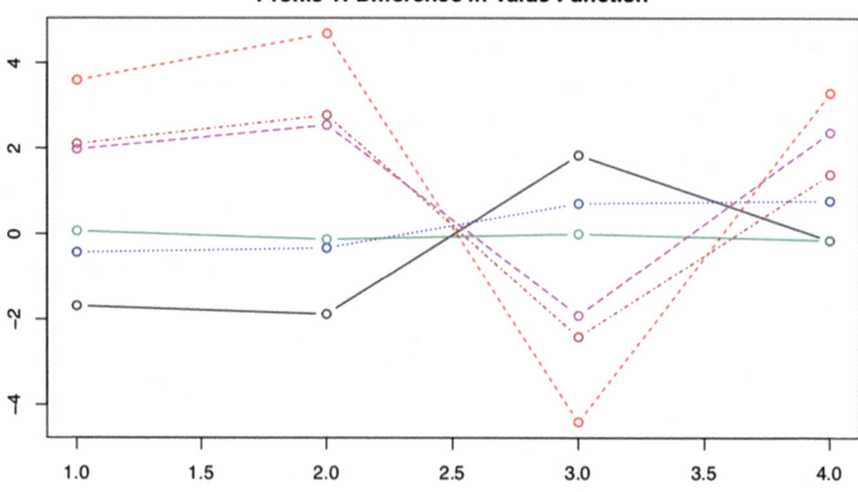

Fig. 5 Expected discounted utility and their differences over time for Profile 1

6 Conclusion

The attribute-level best-worst DCMs considered in this chapter provide a general insight for modeling complex dependencies of time evolving decisions. The decisions for the large data set are guided by a functional form of the expected utility under identifiability constraints. Profile specific trends are displayed and pattern behaviors are exhibited. We highlighted compelling situations that allow shrinkage towards referenced choices and show efficiency data examples to make inferences on the best-worst decisions of interest. Our simulated and aggregated data examples show the flexibility and wide applications of our proposed techniques. Our methodology is easily reproducible. The functional dependency and time evolving structure may accommodate additional arrangements and setups.

A potential area of concern in the application of MDPs for attribute-level best-worst DCEs is the curse of dimensionality as mentioned in Rust (2008). Since the number of attributes, attribute-levels, and profiles grow quickly in the experiment, the estimation process becomes exponentially more difficult. DCEs with larger number of attributes and attribute-levels have more choice sets and pairs to model across time. For discrete processes as is considered in the attribute-level best-worst DCEs, the amount of information that needs to be stored becomes overwhelming. The ability to guide the system becomes difficult due to the increased number of states and choice sets considered. These issues should be considered when using MDPs.

Extensions of this work may include interactions of choice pairs under different correlation structures. The first order Markov dependency structure presented here may be extended to higher order decision processes under stationary and dynamic transition probabilities. Extensions to the continuous time scale case are being explored.

References

Arcidiacono, P., Ellickson, P.B.: Practical methods for estimation of dynamic discrete choice models. Annu. Rev. Econ. **3**(1), 363–394 (2011)

Bellman, R.: The theory of dynamic programming. Bull. Am. Math. Soc **60**(6), 503–515 (1954)

Chadès, I., Chapron, G., Cros, M.-J., Garcia, F., Sabbadin, R.: Mdptoolbox: a multi-platform toolbox to solve stochastic dynamic programming problems. Ecography **37**(9), 916–920 (2014)

Coast, J., Salisbury, C., De Berker, D., Noble, A., Horrocks, S., Peters, T., Flynn, T.: Preferences for aspects of a dermatology consultation. Br. J. Dermatol. **155**(2), 387–392 (2006)

Finn, A., Louviere, J.J.: Determining the appropriate response to evidence of public concern: the case of food safety. J. Public Policy Mark. **11**(2), 12–25 (1992)

Flynn, T.N., Louviere, J.J., Peters, T.J., Coast, J.: Best–worst scaling: what it can do for health care research and how to do it. J. Health Econ. **26**(1), 171–189 (2007)

Flynn, T.N., Louviere, J.J., Marley, A.A., Coast, J., Peters, T.J.: Rescaling quality of life values from discrete choice experiments for use as QALYs: a cautionary tale. Popul. Health Metr. **6**(1), 6 (2008)

Grasshoff, U., Grossmann, H., Holling, H., Schwabe, R.: Optimal paired comparison designs for first-order interactions. Statistics **37**(5), 373–386 (2003)

Grasshoff, U., Grossmann, H., Holling, H., Schwabe, R.: Optimal designs for main effects in linear paired comparison models. J. Stat. Plan. Inference **126**(1), 361–376 (2004)

Grasshoff, U., Grossmann, H., Holling, H., Schwabe, R.: Optimal design for discrete choice experiments. J. Stat. Plan. Inference **143**(1), 167–175 (2013)

Grossmann, H., Grasshoff, U., Schwabe, R.: Approximate and exact optimal designs for paired comparisons of partial profiles when there are two groups of factors. J. Stat. Plan. Inference **139**(3), 1171–1179 (2009)

Hole, A.R.: A discrete choice model with endogenous attribute attendance. Econ. Lett. **110**(3), 203–205 (2011)

Knox, S.A., Viney, R.C., Street, D.J., Haas, M.R., Fiebig, D.G., Weisberg, E., Bateson, D.: What's good and bad about contraceptive products? Pharmacoeconomics **30**(12), 1187–1202 (2012)

Knox, S.A., Viney, R.C., Gu, Y., Hole, A.R., Fiebig, D.G., Street, D.J., Haas, M.R., Weisberg, E., Bateson, D.: The effect of adverse information and positive promotion on women's preferences for prescribed contraceptive products. Soc. Sci. Med. **83**, 70–80 (2013)

Lancsar, E., Louviere, J., Donaldson, C., Currie, G., Burgess, L.: Best worst discrete choice experiments in health: methods and an application. Soc. Sci. Med. **76**, 74–82 (2013)

Lee, T.C., Judge, G., Zellner, A.: Maximum likelihood and Bayesian estimation of transition probabilities. J. Am. Stat. Assoc. **63**(324), 1162–1179 (1968)

Louviere, J., Timmermans, H.: Stated preference and choice models applied to recreation research: a review. Leis. Sci. **12**(1), 9–32 (1990)

Louviere, J.J., Woodworth, G.: Design and analysis of simulated consumer choice or allocation experiments: an approach based on aggregate data. J. Mark. Res. **20**(4), 350–367 (1983)

Louviere, J.J., Woodworth, G.G.: Best-worst scaling: a model for the largest difference judgments. Working Paper. University of Alberta (1991)

Louviere, J.J., Hensher, D.A., Swait, J.D.: Stated Choice Methods: Analysis and Applications. Cambridge University Press, Cambridge (2000)

Louviere, J., Lings, I., Islam, T., Gudergan, S., Flynn, T.: An introduction to the application of (case 1) best–worst scaling in marketing research. Int. J. Res. Mark. **30**(3), 292–303 (2013)

Louviere, J.J., Flynn, T.N., Marley, A.A.J.: Best-Worst Scaling: Theory, Methods and Applications. Cambridge University Press, Cambridge (2015)

Luce, R.D.: On the possible psychophysical laws. Psychol. Rev. **66**(2), 81 (1959)

Marley, A.A., Louviere, J.J.: Some probabilistic models of best, worst, and best–worst choices. J. Math. Psychol. **49**(6), 464–480 (2005)

Marley, A., Pihlens, D.: Models of best–worst choice and ranking among multiattribute options (profiles). J. Math. Psychol. **56**(1), 24–34 (2012)

Marley, A., Flynn, T.N., Louviere, J.: Probabilistic models of set-dependent and attribute-level best–worst choice. J. Math. Psychol. **52**(5), 281–296 (2008)

Marschak, J.: Binary choice constraints on random utility indicators. Technical Report 74 (1960)

McFadden, D.: Conditional logit analysis of qualitative choice behavior. In: Zarembka, P. (ed.) Frontiers in Econometrics, pp. 105–142. Academic Press, New York (1974)

McFadden, D.: Modeling the choice of residential location. Transp. Res. Rec. **673**, 72–77 (1978)

Parvin, S., Wang, P., Uddin, J.: Using best-worst scaling method to examine consumers' value preferences: a multidimensional perspective. Cogent Bus. Manag. **3**(1), 1199110 (2016)

Puterman, M.L.: Markov Decision Processes: Discrete Stochastic Dynamic Programming. Wiley, New York (2014)

Rust, J.: Structural estimation of Markov decision processes. Handb. Econ. **4**, 3081–3143 (1994)

Rust, J.: Dynamic programming. In: Durlauf, S.N., Blume, L.E. (eds.) The New Palgrave Dictionary of Economics. Palgrave Macmillan, Ltd, London (2008)

Stenberg, F., Manca, R., Silvestrov, D.: An algorithmic approach to discrete time non-homogeneous backward semi-Markov reward processes with an application to disability insurance. Methodol. Comput. Appl. Probab. **9**(4), 497–519 (2007)

Street, D.J., Burgess, L.: The Construction of Optimal Stated Choice Experiments: Theory and Methods, vol. 647. Wiley, New York (2007)

Street, D.J., Knox, S.A.: Designing for attribute-level best–worst choice experiments. J. Stat. Theory Pract. 6(2):363–375 (2012)

Thurstone, L.L.: A law of comparative judgment. Psychol. Rev. **34**(4), 273 (1927)

Van Der Pol, M., Currie, G., Kromm, S., Ryan, M.: Specification of the utility function in discrete choice experiments. Value Health **17**(2), 297–301 (2014)

Spatial and Spatio-Temporal Analysis of Precipitation Data from South Carolina

Haigang Liu, David B. Hitchcock, and S. Zahra Samadi

1 Introduction

Spatial and spatio-temporal data are everywhere: we encounter them on TV, in newspapers, on computer screens, on tablets, and on plain paper maps. As a result, researchers in diverse areas are increasingly faced with the task of modeling geographically referenced and temporally correlated data.

The geostatistical analysis of spatial data involves point-referenced data, where $Y(\mathbf{s})$ is a random vector at a location $\mathbf{s} \in \mathcal{R}^r$, where $\mathbf{s}$ varies continuously over D, a fixed subset of $\mathcal{R}^r$ that contains an r-dimensional rectangle of positive volume (Banerjee et al. 2014). The sample points are measurements of some phenomenon such as precipitation measurements from meteorological stations or elevation heights. The geostatistical analysis models a surface using the values from the measured locations to predict values for each location in the landscape.

Spatial statistics methods have been frequently used in applied statistics as well as water resources engineering. The work of Thiessen (1911) was the first attempt in using interpolation methods in hydrology. Sharon (1972) used an average of the observations from a number of rain gages to obtain estimates of the areal rainfall. Soon after, Benzécri (1973), Delfiner and Delhomme (1975), and Delhomme (1978) applied the various geostatistical methods such as variograms and kriging methods in modeling rainfall. The work of Troutman (1983), Tabios

H. Liu · D. B. Hitchcock (✉)
Department of Statistics, University of South Carolina, Columbia, SC, USA
e-mail: haigang@email.sc.edu; hitchcock@stat.sc.edu

S. Z. Samadi
Department of Civil and Environmental Engineering, University of South Carolina, Columbia, SC, USA
e-mail: samadi@cec.sc.edu

© Springer Nature Switzerland AG 2019

N. Diawara (ed.), *Modern Statistical Methods for Spatial and Multivariate Data*, STEAM-H: Science, Technology, Engineering, Agriculture, Mathematics & Health, https://doi.org/10.1007/978-3-030-11431-2_2

and Salas (1985), Georgakakos and Kavvas (1987), Isaaks and Srivastava (1989), Kumar and Foufoula-Georgiou (1994), Deidda (2000), Ferraris et al. (2003), Ciach and Krajewski (2006), Berne et al. (2009), Ly et al. (2011), and Dumitrescu et al. (2016) further advanced the application of geostatistical methods in rainfall prediction. The theoretical basis of the geostatistical approach was strengthened using Bayesian inference via the Markov Chain Monte Carlo (MCMC) algorithm introduced by Metropolis et al. (1953). MCMC was subsequently adapted by Hastings (1970) for statistical problems and further applied by Diggle et al. (1998) in geostatistical studies. Recent developments in MCMC computing now allow fully Bayesian analyses of sophisticated multilevel models for complex geographically referenced data. This approach also offers full inference for non-Gaussian spatial data, multivariate spatial data, spatio-temporal data, and solutions to problems such as geographic and temporal misalignment of spatial data layers (Banerjee et al. 2014).

The data we are studying are monthly rainfall data measured across the state of South Carolina from the start of 2011 to the end of 2015. The precipitation record in 2015 is of particular interest because a storm in October 2015 in North America triggered a high precipitation event, which caused historic flash flooding across North and South Carolina. Rainfall across parts of South Carolina reached 500-year-event levels (NBC News, October 4, 2015). Accumulations reached 24.23 in. near Boone Hall (Mount Pleasant, Charleston County) by 11:00 a.m. Eastern Time on October 4, 2015. Charleston International Airport saw a record 24-h rainfall of 11.5 in. (290 mm) on October 3 (Santorelli, October 4, 2015). Some areas experienced more than 20 in. of rainfall over the 5-day period. Many locations recorded rainfall rates of 2 in. per hour (National Oceanic and Atmospheric Administration (NOAA), U.S. Department of Commerce, 2015).

The extraordinary rainfall event was generated by the movement of very moist air over a stalled frontal boundary near the coast. The clockwise circulation around a stalled upper level low over southern Georgia directed a narrow plume of tropical moisture northward and then westward across the Carolinas over the course of 4 days. A low pressure system off the US southeast coast, as well as tropical moisture related to Hurricane Joaquin (a category 4 hurricane) was the underlying meteorological cause of the record rainfall over South Carolina during October 1–5, 2015 (NOAA, U.S. Department of Commerce 2015).

Flooding from this event resulted in 19 fatalities, according to the South Carolina Emergency Management Department, and South Carolina state officials said damage losses were 1.492 billion dollars (NOAA, U.S. Department of Commerce 2015). The heavy rainfall and floods, combined with aging and inadequate drainage infrastructure, resulted in the failure of many dams and flooding of many roads, bridges, and conveyance facilities, thereby causing extremely dangerous and life-threatening situations.

The chapter is arranged as follows: in Sect. 2, we give an overview of our precipitation data, in conjunction with some other variables, e.g., sea surface temperature, which might help explain the behavior of the precipitation. In Sect. 3, we introduce the kriging method to analyze the precipitation using a pure spatial analysis. In Sect. 4, some methods in seasonal trend removal are discussed. In Sect. 5, the Gaussian process is introduced to build a spatio-temporal model.

2 Data Description

2.1 Overview

The original data used in this research are the daily precipitation records in South Carolina from National Oceanic and Atmosphere Administration (NOAA) between 2011 and 2015. The original data files include daily precipitation, maximum temperature, and minimum temperature, along with the latitude, longitude, and elevation of each observation's location.

In addition, to investigate the effect of El Niño-Southern Oscillation (ENSO) activity on precipitation, we have calculated an index based on the monthly sea surface temperature (SST). The derivation of our index is given in Sect. 2.3.

2.2 Data Preprocessing

We collected 281 unique meteorological locations in South Carolina with varying completeness of data. For instance, if we look at the most recent 5 years (2011–2015), 31 locations do not have any record of precipitation while 65 locations have a complete record. The other 185 locations contain missing data ranging from 30% to less than 5% of the total data set size.

In Fig. 1, we plot all the meteorological locations with an available precipitation record on October 3, 2015, when the storm struck South Carolina. Note that smoothing is necessary since most of observations are clustered in several regions. See Bivand et al. (2008) for more information about the sp package, which provides a comprehensive solution for spatial data visualization.

We aggregate the daily records into monthly variables. The monthly maximum of precipitation is calculated since we are interested in capturing the extreme rainfall behavior which might lead to flooding subsequently. The monthly midrange temperature, which reflects the general warmth of that month, is computed by averaging the highest and the lowest daily temperature for that month.

To incorporate more temperature information, we find the range of daily maxima over a month. We similarly obtain the range of the daily minima. Lastly, for each location, we also find an overall range, the difference of the maximum and minimum temperature of that month.

In the data set, several variables, e.g., precipitation, elevation, and temperature have missing values. We replace each missing observation with the weighted average of its neighbors. The weights are determined by the distance between

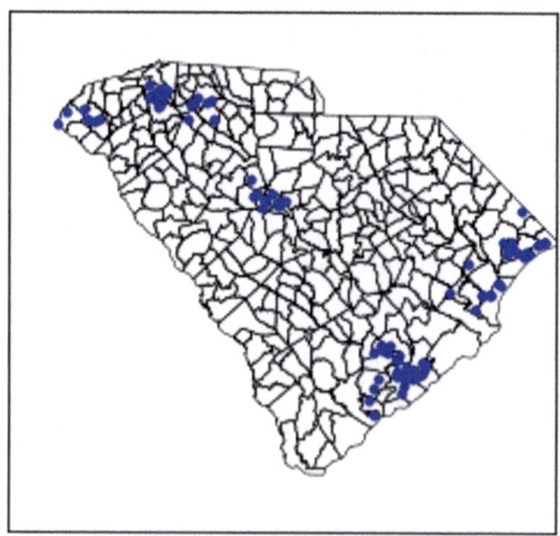

Fig. 1 The meteorological locations with available record of precipitation on October 3, 2015

locations. In other words, if we denote the missing value at $\mathbf{s}^*$ by $Y(\mathbf{s}^*)$, then $\sum_{i=1}^{n} w(\mathbf{s}_i) Y(\mathbf{s}_i)$ can be used as the imputed value, where

$$w(\mathbf{s}_i) = K\left(\frac{||\mathbf{s}^* - \mathbf{s}_i||}{h}\right) \bigg/ \sum_{i=1}^{n} K\left(\frac{||\mathbf{s}^* - \mathbf{s}_i||}{h}\right). \tag{1}$$

Note that $||\mathbf{s}_i - \mathbf{s}^*||$ refers to the haversine distance rather than the Euclidean distance. We impute missing data based on neighboring observations because doing so takes the spatial correlation into consideration.

2.3 A Sea Surface Temperature (SST)-Related Variable

El Niño-Southern Oscillation (ENSO) is an irregular variation in winds and sea surface temperature (SST) over the tropical eastern Pacific Ocean, affecting much of the tropic and subtropics. Like other climate indices, ENSO occurs irregularly and is associated with changing in physical pattern of temperature and precipitation. Figure 2 gives the plot of sea surface temperature for ocean locations off the coast of South Carolina in June 2015. In this figure, dark colors correspond to cooler sea temperature values. Scientists believe that the ENSO has a significant influence on precipitation and hence controls flood magnitude and frequency. We thus include an SST-based index as a proxy for the ENSO activity. Since our rainfall data are observed for inland locations, we must define our index related to SST for such inland locations, rather than for off-shore locations where sea temperature is actually measured.

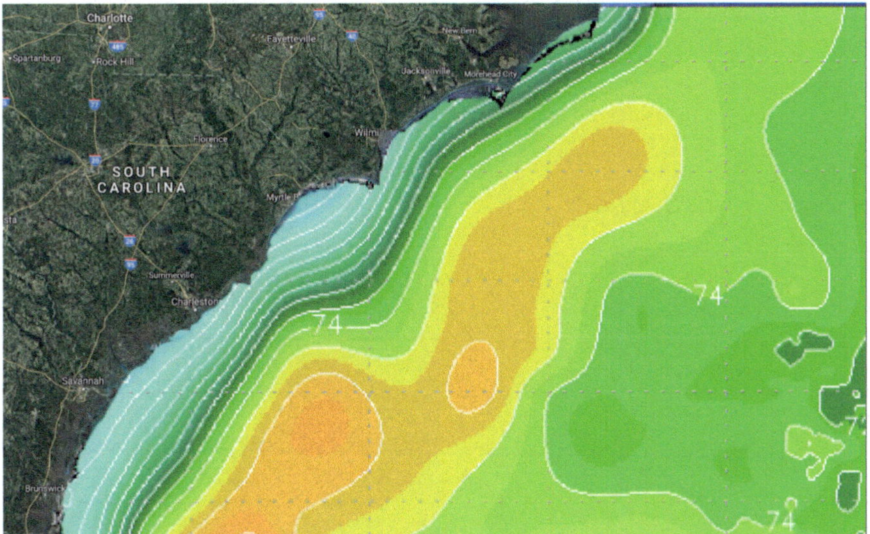

Fig. 2 The sea surface temperature near South Carolina

For any inland location $\mathbf{s}_i$ at a given month, we build an index based on the SST values of the nearest n adjacent ocean observation points $\{\mathbf{z}_j\}$, where $j = 1, \ldots, n$. Denote this SST-based index as $W(\mathbf{s}_i)$ for the ith inland location. It follows that

$$W(\mathbf{s}_i) = \frac{1}{n} \sum_{j=1}^{n} \left(\frac{w_j}{\sum_{l=1}^{n} w_l} \right) \mathrm{SST}(\mathbf{z}_j), \tag{2}$$

where the weight w_j can be determined by the kernel function $K(\|\mathbf{s}_i - \mathbf{z}_j\|)$ for $j = 1, \ldots, n$, which is symmetric around 0. We use the standard normal density as the kernel function. The kernel function includes a bandwidth h, thus making $w_j = \frac{1}{h} K(\frac{\|\mathbf{s}_i - \mathbf{z}_j\|}{h})$. The bandwidth parameter h is set to 0.25 times the range of all of the distances.

Additionally, we simplify the calculation by considering only locations within a certain threshold. Figure 3 gives a demonstration to calculate the SST-related index for Columbia, South Carolina. We first determine the sea temperature records to be included based on a 300-mile threshold. For the included measurements, we find their weights by calculating their distance to Columbia, and derive the SST-related index based on (2). Note that the closer a location is to the coast, the more sea surface temperature records are used to derive an SST-related index for that location.

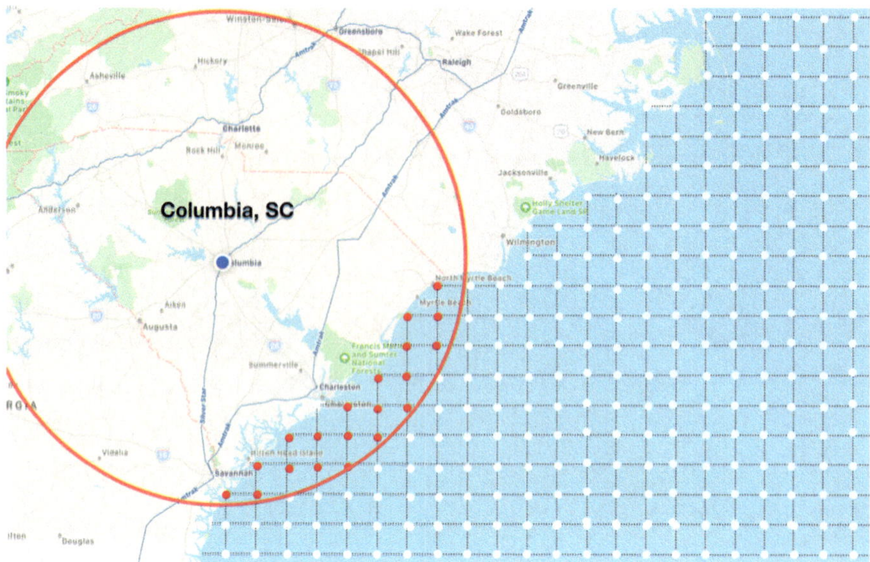

Fig. 3 A demonstration of the calculation of the SST-related variable. The red points are the observations that are included in the calculation

3 Precipitation Modeling: A Spatial Perspective

In this section, we use a spatial model for the rainfall data without considering the temporal aspect. Since geostatistical data feature a strong correlation between adjacent locations, we start by modeling the covariance structure with a variogram, and then we propose two methods of predicting the rainfall for new location.

3.1 Describing the Spatial Structure: Variogram

We assume that our spatial process has a mean, $\mu(\mathbf{s}) = E(Y(\mathbf{s}))$, and that the variance of $Y(\mathbf{s})$ exists for all $\mathbf{s} \in D$. The process $Y(\mathbf{s})$ is said to be Gaussian if, for any $n \geq 1$ and any set of sites $\{\mathbf{s}_1 \ldots, \mathbf{s}_n\}$, $\mathbf{Y} = (Y(\mathbf{s}_1), \ldots, Y(\mathbf{s}_n))^T$ has a multivariate normal distribution. Moreover, the process is *intrinsic stationary* if, for any given $n \geq 1$, any set of n sites $\{\mathbf{s}_1, \ldots, \mathbf{s}_n\}$ and any $h \in \mathcal{R}^r$, we have $E[Y(\mathbf{s}+\mathbf{h}) - Y(\mathbf{s})] = 0$, and $E[Y(\mathbf{s}+\mathbf{h}) - Y(\mathbf{s})]^2 = \text{Var}(Y(\mathbf{s}+\mathbf{h}) - Y(\mathbf{s})) = 2\gamma(\mathbf{h})$ (Banerjee et al. 2014).

In other words, $E[Y(\mathbf{s} + \mathbf{h}) - Y(\mathbf{s})]^2$ only depends on $\mathbf{h}$, and not the particular choice of $\mathbf{s}$. The function $2\gamma(\mathbf{h})$ is then called the *variogram*, and $\gamma(\mathbf{h})$ is called the *semivariogram*. Another important concept is that of an *isotropic* variogram. If the semivariogram function $\gamma(\mathbf{h})$ depends upon the separation vector only through its length $\|\mathbf{h}\|$ (distance between observations), then the variogram is isotropic. Otherwise, it is *anisotropic*. Isotropic variograms are popular because of simplicity, interpretability, and, in particular, because a number of relatively simple parametric forms are available as candidates for the semivariogram, e.g., linear, exponential, Gaussian, or Matérn (or K-Bessel).

A variogram model is chosen by plotting the empirical semivariogram, a simple nonparametric estimate of the semivariogram, and then comparing it to the various theoretical parametric forms (Matheron 1963). For demonstration purposes, we choose the precipitation values of October 13 in 2015, shortly after the flood struck South Carolina. Assuming intrinsic stationarity and isotropy, the Matérn model is used due to its better fit to the empirical semivariogram. The correlation function of this model allows control of spatial association and smoothness. See Fig. 4 for a plot of this fit.

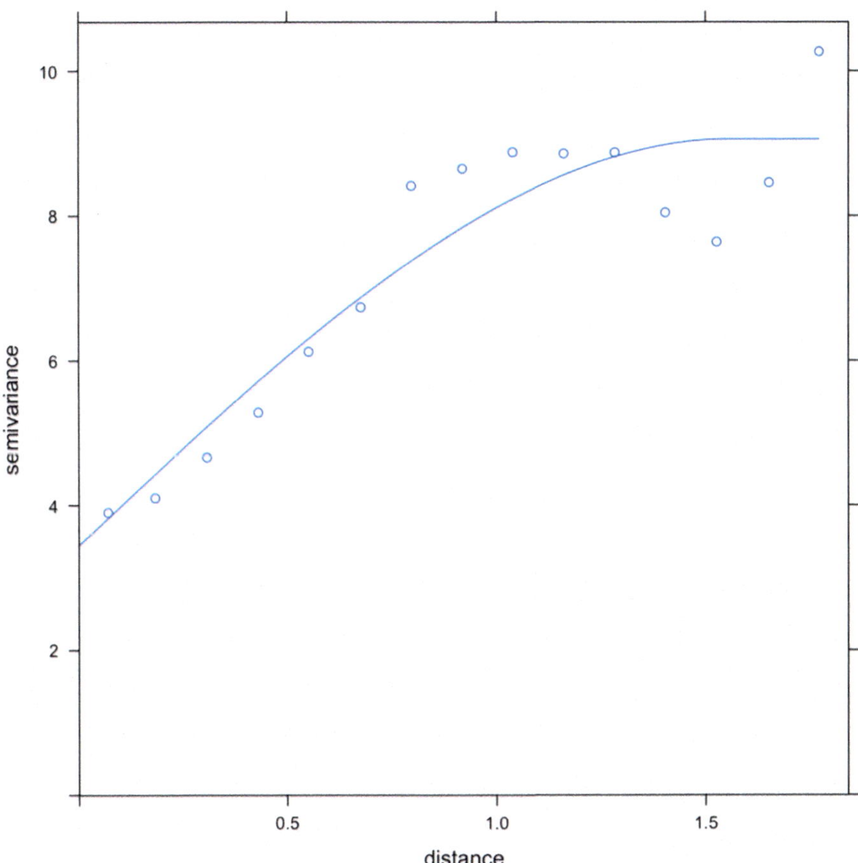

Fig. 4 The empirical and parametric (Matérn) variogram for the precipitation values in October 13, 2015

3.2 Spatial Prediction

Inverse Distance Weighted Interpolation

We use inverse distance weighting (IDW) (Bivand et al. 2008) to compute a spatially continuous rainfall estimate as a weighted average for a given location $\mathbf{s}_0$,

$$\hat{Z}(\mathbf{s}_0) = \frac{\sum w(\mathbf{s}_i)Z(\mathbf{s}_i)}{\sum w(\mathbf{s}_i)}, \quad \text{where } w(\mathbf{s}_i) = ||\mathbf{s}_i - \mathbf{s}_0||^{-p}.$$

In other words, the weight of a given observed location is based on its L_p-distance to the interpolation location. If location $\mathbf{s}_0$ happens to have an observation, then the observation itself will be used to avoid the case of infinite weights. The weight assigned to data points will be more influenced by neighboring points when they are more clustered. The best p found by cross validation for the analysis of our data set is approximately 2.5.

Although this method does not incorporate the covariates, it still possesses some desirable features. For instance, we can make a prediction for the rainfall amount at every single location with a latitude and longitude.

Linear Gaussian Process Model (Kriging)

Since our precipitation data in the study are geostatistical data, we may employ a linear Gaussian process model (Cressie 1993). We start by defining the spatial process at location $\mathbf{s} \in \mathcal{R}^d$ as

$$Z(\mathbf{s}) = X(\mathbf{s})\boldsymbol{\beta} + w(\mathbf{s}), \tag{3}$$

where $\mathbf{X}(\mathbf{s})$ is a set of p covariates associated with each site $\mathbf{s}$, and $\boldsymbol{\beta}$ is a p-dimensional vector of coefficients. Spatial dependence is imposed via the residual terms, i.e., $w(\mathbf{s})$. Specifically, we model $\{w(\mathbf{s}) : \mathbf{s} \in \mathcal{R}^d\}$ as a zero mean Gaussian process. In other words, the vector $\mathbf{w} = (w(\mathbf{s}_1), \ldots, w(\mathbf{s}_n))^T$ follows $\mathbf{w}|\Theta \sim N_n(\mathbf{0}, \boldsymbol{\Sigma}(\Theta))$. We assume $\boldsymbol{\Sigma}$ to be a symmetric and positive definite matrix in order to end up with a sensible distribution. To ensure these conditions, $\boldsymbol{\Sigma}(\Theta)$ can be treated as a function of Θ with certain constraints, which are tantamount to specifying a variogram model.

Among several variogram structures, e.g., spherical, Gaussian, exponential, etc. we choose the exponential covariance with parameters $\Theta = (\psi, \kappa, \phi)$, where $\psi, \kappa, \phi > 0$. The exponential covariance $\boldsymbol{\Sigma}(\Theta)$ has the form

$$\boldsymbol{\Sigma}(\Theta) = \psi \mathbf{I} + \kappa H(\phi), \quad \text{where } H(\phi) = \exp(-||\mathbf{s}_i - \mathbf{s}_j||)/\phi).$$

Note that $||\mathbf{s}_i - \mathbf{s}_j||$ is the Euclidean distance between location i and j. Another type of distance, *Geodesic*, takes the curvature of the earth's surface into consideration.

We use Euclidean distance since most of our distances are between South Carolina counties and the effects of curvature are thus negligible.

The exponential model enjoys a simple interpretation. The "nugget" in a variogram graph is represented by ψ in this model, and this nugget is also the variance of the non-spatial error. Moreover, κ and ϕ dictate the scale and range of the spatial dependence, respectively. Also note that the exponential model assumes the covariance and hence dependence between two locations decreases as distance between locations increases, which is sensible for the study of rainfall behavior.

Letting $\mathbf{Z} = (Z(\mathbf{s}_1), \ldots, Z(\mathbf{s}_n))^T$, we estimate the multivariate normal distribution for $\mathbf{Z}$ after parameter estimation. To find the unknown parameters Θ and $\boldsymbol{\beta}$, we use Bayesian methods implemented by the `spTimer` package in R (Bakar and Sahu 2015), which requires users to provide sensible prior information based on sample variogram graphs. Note that this model fitting process will collapse if we start with initial values far from the true value.

Monte Carlo Simulation for Kriging

Predictions of the process, $\mathbf{Z}^* = (Z(\mathbf{s}_1^*), \ldots, Z(\mathbf{s}_m^*))^T$, where $\mathbf{s}_i^*$ is the ith new location, can be obtained via the posterior predictive distribution

$$\pi(\mathbf{Z}^*|\mathbf{Z}) = \int \pi(\mathbf{Z}^*|\mathbf{Z}, \Theta, \boldsymbol{\beta})\pi(\Theta, \boldsymbol{\beta}|\mathbf{Z})d\Theta d\boldsymbol{\beta},$$

by sampling from the posterior predictive distribution in two steps:

- Step 1: Simulate $\Theta', \boldsymbol{\beta}' \sim \pi(\Theta, \boldsymbol{\beta}|\mathbf{Z})$ by the Metropolis–Hastings algorithm.
- Step 2: Simulate $\mathbf{Z}^*|\Theta', \boldsymbol{\beta}, \mathbf{Z}$ from a multivariate normal density.

For step 1, it suffices to find the posterior distribution $\pi(\Theta, \boldsymbol{\beta}|\mathbf{Z})$ based on (1) and (2). The posterior distribution has low dimension as long as we do not have many covariates. The major challenge is that since covariance parameters might be highly correlated, one must expect autocorrelation issues in the sampler, which can be alleviated by a block updating scheme, a scheme that generates multiple covariance parameters in a single Metropolis–Hastings step.

For step 2, the joint distribution of $\mathbf{Z}$ and $\mathbf{Z}^*$ is given by

$$\begin{bmatrix} \mathbf{Z} \\ \mathbf{Z}^* \end{bmatrix} |\Theta, \boldsymbol{\beta} \sim N\left(\begin{bmatrix} \mu_1 \\ \mu_2 \end{bmatrix}, \begin{bmatrix} \Sigma_{11} & \Sigma_{12} \\ \Sigma_{21} & \Sigma_{22} \end{bmatrix}\right)$$

based on which one can find the conditional distribution of $\mathbf{Z}^*|\Theta', \boldsymbol{\beta}, \mathbf{Z}$. According to Anderson (2003), it follows that

$$E(\mathbf{Z}^*|\Theta', \boldsymbol{\beta}, \mathbf{Z}) = \mu_2 + \Sigma_{21}\Sigma_{11}^{-1}(\mathbf{Z} - \mu_1),$$

$$\text{Var}(\mathbf{Z}^*|\Theta', \boldsymbol{\beta}, \mathbf{Z}) = \Sigma_{22} - \Sigma_{21}\Sigma_{11}^{-1}\Sigma_{12}.$$

Hence, one can obtain simulated observations that follow a given covariance structure by iterating between step 1 and step 2. Bivand et al. (2008) suggest the method of sequential simulation: (1) compute the conditional distribution with our given data, (2) draw a value from this conditional distribution, (3) add this value into the data set, and (4) repeat steps (1)–(3).

As $\mathbf{Z}$ becomes a larger matrix as more data are generated, the algorithm becomes more and more expensive. Many strategies are proposed for reducing the considerable computational burden posed by matrix operations, including the use of covariance functions (Hughes and Haran 2013) as well as setting a maximum number of neighbors (Bivand et al. 2008). In our study, we used the maximum number of neighbors with the nearest 40 observations.

We illustrate prediction by modeling rainfall in South Carolina on October 13, 2015 with a kriging model that assumes an exponential spatial covariance structure. Using the Monte Carlo approach described above, we predict by simulating from the posterior predictive distribution. This can be done repeatedly to give a sense of the variability associated with the spatial predictions. Figure 5 demonstrates ten simu-

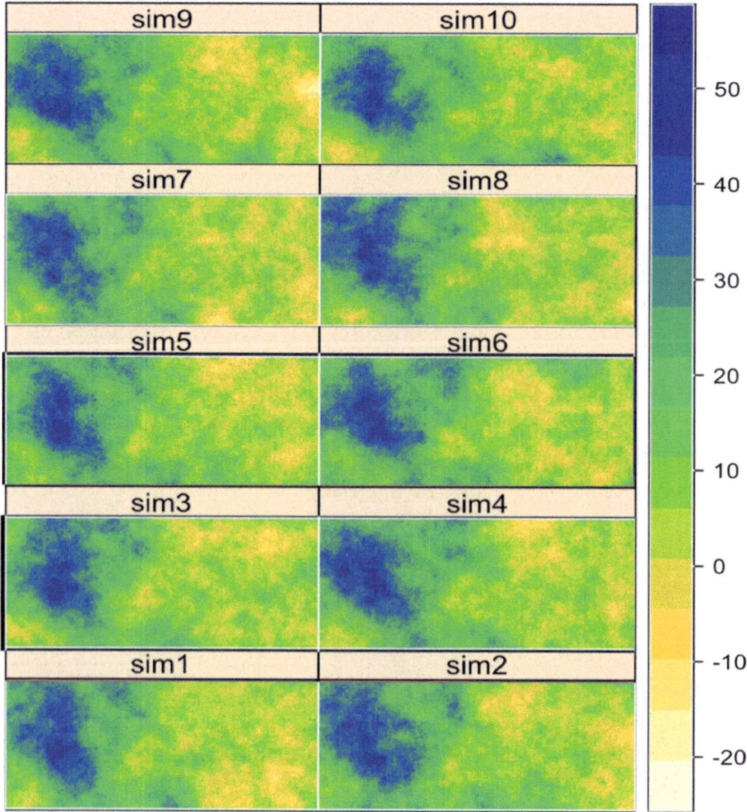

Fig. 5 Ten simulated precipitation heat maps based on kriging. The darker color indicates heavier precipitation and vice versa. A consistent look reveals a robust performance of the kriging model

lated predictions of the spatial distribution of rainfall amounts in a small rectangular spatial area in the northwest corner of South Carolina. The darker color indicates heavier predicted precipitation and the lighter color a small predicted rainfall. The consistent pattern across all ten simulations reveals a robust performance of the kriging model. A pointwise prediction at any spatial location could be obtained by averaging the predicted rainfall values at that location across all ten simulations.

4 Seasonal Trend Removal

We now analyze the geostatistical rainfall data across time. Due to the nature of our rainfall data, the seasonality is of particular interest when we model the temporal trend. We propose two methods to remove the seasonal trend in this section.

4.1 Harmonic Regression

To remove the seasonal trend, one approach is to fit a first-order harmonic regression model with terms $\sin(x)$ and $\cos(x)$. In addition, we set $x = 2\pi t$ if the period is 1. In our case, it is justifiable to set the period as 12 since the monthly rainfall is measured, and thus $x = (\pi/6)t$ is used. Hence, one can regress the precipitation y against dependent variables $\sin((\pi/6)\,t)$ and $\cos((\pi/6)\,t)$. The omnibus F-test to test for the usefulness of the trigonometric terms in this multiple regression model gives a p-value close to 1, which confirms the existence of seasonality.

One can also use a second-order harmonic model to capture more complex behavior, in which two more terms, $\sin[(4\pi/\omega)t]$ and $\cos[(4\pi/\omega)t]$ are included, where ω is the periodic parameter. However, for our rainfall data, it is unnecessary to include these two other terms since we observe no great improvement in model fit by introducing the extra terms (see Fig. 6).

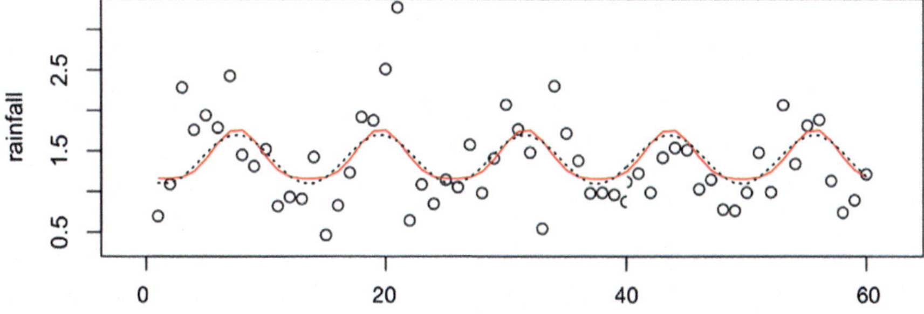

Fig. 6 The fitted model based on the first- and second-order harmonic models. The dotted line corresponds to the second-order model, and the solid red line corresponds to the first-order model

4.2   Seasonality Indicator

Another approach to model seasonality in the spatio-temporal model is the seasonal means model. Specifically, one indicator variable will be 1 if the record is collected from January to March, and will be 0 otherwise. Similarly, another dummy variable indicates the month April to June while a third dummy variable indicates July to September. Lastly, if all three variables are 0, then the observation is from the last 3 months of the year. Note that one could also include dummy variables for months in a similar way if necessary, but we have found that it is sufficient to model the means of the four seasons.

5   Precipitation Modeling: A Spatio-Temporal Perspective

In this section, we discuss how to model spatio-temporal data with two different methods, the Gaussian process (GP) model and autoregressive (AR) model. The latter model is an extension of the Gaussian process model obtained by introducing an autoregressive term.

5.1   Gaussian Process (GP) Model

The independent Gaussian process (GP) model (Cressie and Wikle 2015; Gelfand et al. 2010) is specified hierarchically in two stages,

$$\mathbf{Z}_t = \boldsymbol{\mu}_t + \boldsymbol{\epsilon}_t \tag{4}$$

$$\boldsymbol{\mu}_t = \mathbf{X}_t \boldsymbol{\beta} + \boldsymbol{\eta}_t, \tag{5}$$

in which $\mathbf{Z}_t = (Z(\mathbf{s}_1, t), \ldots, Z(\mathbf{s}_n, t))^T$, which defines the response variable for all n locations at time t. It is known that $\mathbf{s}_1, \ldots, \mathbf{s}_n$ can be indexed by latitude and longitude. In the first layer, $\mathbf{Z}_t$ is defined by a simple mean model plus a pure white noise term, $\boldsymbol{\epsilon}_t$. We therefore assume that

$$\boldsymbol{\epsilon}_t \sim N(\mathbf{0}, \sigma_\epsilon^2 \mathbf{I}_n), \tag{6}$$

where the σ_ϵ^2 is the pure error variance and $\mathbf{I}_n$ is the identity matrix.

The second level models $\boldsymbol{\mu}_t$ as the sum of fixed covariates and random effects at time t. The fixed term, $\mathbf{X}_t \boldsymbol{\beta}$, comes from the covariates, and $\boldsymbol{\eta}_t$ is the spatio-temporal random effects, $\boldsymbol{\eta}_t = (\eta(\mathbf{s}_1, t), \ldots, \eta(\mathbf{s}_n, t))^T$. Similar to $\boldsymbol{\epsilon}_t$, $\boldsymbol{\eta}_t$ also follows a multivariate normal distribution whose mean vector is $\mathbf{0}$. However, $\boldsymbol{\eta}_t$ has a more complicated covariance matrix than does $\boldsymbol{\epsilon}_t$.

We use the exponential function to specify the correlation matrix of the random effects. The correlation strength is solely based on the distance between $\mathbf{s}_i$ and $\mathbf{s}_j$, which is given by

$$\boldsymbol{\Sigma}_\eta = \sigma_\eta^2 H(\phi) + \tau^2 \mathbf{I}_n,$$

where $H(\phi) = \exp(-||\mathbf{s}_i - \mathbf{s}_j||)/\phi)$, and $||\mathbf{s}_i - \mathbf{s}_j||$ indicates the spatial distance between location i and j. This function is used to determine each element in the matrix $\mathbf{S}_\eta$, where $\boldsymbol{\Sigma}_\eta = \sigma_\eta^2 \mathbf{S}_\eta$. This parameterization allows σ_η^2 to capture the invariant spatial variance, and $\mathbf{S}_\eta$ is used to capture the spatial correlation.

The posterior distribution involves three layers, i.e., the prior distribution for parameters, the mean model, and the random effects model. We will set aside the prior for later discussion and use $\pi(\boldsymbol{\theta}) = \pi(\boldsymbol{\beta}, v, \phi, \sigma_\eta^2, \sigma_\epsilon^2)$ to refer to the prior in general. Thus the posterior is given by

$$g(\boldsymbol{\theta}, \boldsymbol{\mu}|\mathbf{Z}) = \pi(\boldsymbol{\theta}) \times \prod_{t=1}^{N} f_n(\mathbf{Z}_t|\boldsymbol{\mu}_t, \sigma_\epsilon^2) g_n(\boldsymbol{\mu}_t|\boldsymbol{\beta}, v, \phi, \sigma_\eta^2). \tag{7}$$

To be specific, we use $f_n(\cdot)$ and $g_n(\cdot)$ to indicate an n-dimensional distribution function. In this case, each of them is a multivariate normal distribution, and n is the number of locations in the data set and N is the number of time points. $\boldsymbol{\mu}_t$ is the vector of random effects for time t and we use $\boldsymbol{\mu}$ on the left-hand side to refer to the collection of all random effects.

Since both $\mathbf{Z}_t$ and $\boldsymbol{\mu}_t$ follow a multivariate normal distribution, their density functions are given as follows:

$$f_n(\mathbf{Z}_t|\boldsymbol{\mu}_t, \sigma_\epsilon^2) = \frac{1}{\sqrt{(2\pi)^n |\sigma_\epsilon^2 \mathbf{I}_n|}} \exp\left(-\frac{1}{2\sigma_\epsilon^2}(\mathbf{Z}_t - \boldsymbol{\mu}_t)^T (\mathbf{Z}_t - \boldsymbol{\mu}_t)\right), \tag{8}$$

$$g_n(\boldsymbol{\mu}_t|\mathbf{S}_\eta, \sigma_\eta^2, \boldsymbol{\beta}) = \frac{1}{\sqrt{(2\pi)^n |\sigma_\eta^2 \mathbf{S}_\eta|}} \exp\left(-\frac{1}{2\sigma_\eta^2}(\boldsymbol{\mu}_t - \mathbf{X}_t\boldsymbol{\beta})^T S_\eta^{-1}(\boldsymbol{\mu}_t - \mathbf{X}_t\boldsymbol{\beta})\right),$$
$$\tag{9}$$

Thus the posterior distribution is given by plugging (8) and (9) into (7). The logarithm of the joint posterior distribution of the parameters for this Gaussian process model is given by

$$\log \pi(\sigma_\epsilon^2, \sigma_\eta^2, \boldsymbol{\mu}, \boldsymbol{\beta}, v, \phi|\mathbf{Z}) \propto \frac{N}{2} \log \sigma_\epsilon^2 - \frac{1}{2\sigma_\epsilon^2} \sum_{t=1}^{N} (\mathbf{Z}_t - \boldsymbol{\mu}_t)^T (\mathbf{Z}_t - \boldsymbol{\mu}_t)$$

$$- \frac{N}{2} \log |\sigma_\eta^2 \mathbf{S}_\eta| - \frac{1}{2\sigma_\eta^2} \sum_{i=1}^{N} \left(-\frac{1}{2\sigma_\eta^2}(\boldsymbol{\mu}_t - \mathbf{X}_t\boldsymbol{\beta})^T S_\eta^{-1}(\boldsymbol{\mu}_t - \mathbf{X}_t\boldsymbol{\beta})\right) + \log \pi(\boldsymbol{\theta}).$$

We specify the prior $\pi(\boldsymbol{\theta})$ to reflect the assumption that $\boldsymbol{\beta}$, ν, ϕ, σ_η^2, and σ_ϵ^2 are mutually independent, so the joint prior is the product of the marginal prior densities, which are given as follows: All the parameters describing the mean, e.g., $\boldsymbol{\beta}$ and ρ (see Sect. 5.2) are given independent normal prior distributions, with the prior on ρ truncated to have support on $(-1, 1)$. We assume ϕ and ν both follow uniform distributions, while the prior for the precision (inverse of variance) parameter is a gamma distribution. We choose the hyperparameters to make these prior distributions very diffuse.

5.2 Autoregressive (AR) Model

In this section, we introduce the autoregressive model (Sahu and Bakar 2012). The hierarchical AR(1) model is given as follows:

$$\mathbf{Z}_t = \boldsymbol{\mu}_t + \boldsymbol{\epsilon}_t$$

$$\boldsymbol{\mu}_t = \rho\boldsymbol{\mu}_{t-1} + \mathbf{X}_t\boldsymbol{\beta} + \boldsymbol{\eta}_t,$$

where ρ denotes the unknown temporal correlation parameter assumed to be in the interval $(-1, 1)$. Obviously, for $\rho = 0$, these models reduce to the GP model described in Sect. 5.1.

The autoregressive model requires specification of the initial term, the first random effect, which has mean $\boldsymbol{\beta}_0$ and covariance matrix $\sigma_0^2 \mathbf{S}_0$. The correlation matrix $\mathbf{S}_0$ is obtained using the exponential correlation function. The derivation of the posterior distribution is similar to that in GP model with $\rho = 0$. The logarithm of the posterior distribution of the parameters is now given by

$$\log\pi(\sigma_\epsilon^2, \sigma_\eta^2, \boldsymbol{\mu}, \boldsymbol{\beta}, \upsilon, \phi | \mathbf{Z}) \propto \frac{N}{2}\log\sigma_\epsilon^2 - \frac{1}{2\sigma_\epsilon^2}\sum_{t=1}^{N}(\mathbf{Z}_t - \boldsymbol{\mu}_t)^T(\mathbf{Z}_t - \boldsymbol{\mu}_t)$$

$$-\frac{N}{2}\log|\sigma_\eta^2 \mathbf{S}_\eta|$$

$$-\frac{1}{2\sigma_\eta^2}\sum_{i=1}^{N}\left(-\frac{1}{2\sigma_\eta^2}(\boldsymbol{\mu}_t - \rho\boldsymbol{\mu}_{t-1} - \mathbf{X}_t\boldsymbol{\beta})^T \mathbf{S}_\eta^{-1}(\boldsymbol{\mu}_t - \rho\boldsymbol{\mu}_{t-1} - \mathbf{X}_t\boldsymbol{\beta})\right)$$

$$-\frac{1}{2}\log|\sigma_0^2 \mathbf{S}_0| - \frac{1}{2\sigma_0^2}(\boldsymbol{\mu}_0 - \boldsymbol{\beta}_0)^T \mathbf{S}_0^{-1}(\boldsymbol{\mu}_0 - \boldsymbol{\beta}_0) + \log\pi(\boldsymbol{\theta})$$

Note that $\boldsymbol{\beta}_0$ is only a mean vector for the initial random effect term, which is different from $\boldsymbol{\beta}$, which refers to regression coefficients corresponding to covariates $\mathbf{X}$. In other words, the terms in the last line (except $\log\pi(\boldsymbol{\theta})$) derive from the initial random effect term.

5.3 *Model Fitting*

In this section, we fit the AR(1) model with monthly precipitation data from the beginning of year 2011 to the end of year 2015. A natural log transformation was initially applied to the precipitation to improve the model fit and ensure positive predicted rainfall values once we back-transform by exponentiating the predicted log-rainfall values. We include temperature range, sea surface temperature, and elevation as monthly covariates.

We initially found that ordinary temperature measurements such as the monthly average temperature were not apparently related to precipitation after accounting for the season and thus we did not include these in the model. However, measurements of variability in temperature over each month, e.g., the range of daily maxima and the range of daily minima over a month, were believed to have an effect on precipitation and thus we include these to determine whether their effects are significant.

We also include a flood-year indicator as a dummy variable, where data from 2015 is labeled as 1 and otherwise 0, to account for the unusual October precipitation amounts in this year. Interaction terms involving the dummy variable were also tested, none of which were statistically significant and were thus removed from the final model. The acceptance rate from Metropolis step for all parameters is 42.97% and a brief summary of model fitting details is given as follows:

```
------------------------------------------------------------
Model: AR
Call: LOG ~ RANGE_OVERALL + RANGE_LOW  + RANGE_HIGH
+ SST + ELEVATION + SST * RANG E_LOW + Year2015

Iterations: 5000
nBurn: 1000
Acceptance rate: 29.76
------------------------------------------------------------
Parameters
                  Mean  Median      SD Low2.5p Up97.5p
(Intercept)     0.3635  0.3689  0.1363  0.0894  0.6265
RANGE_OVERALL  -0.0006 -0.0006  0.0017 -0.0039  0.0027
RANGE_LOW       0.0017  0.0017  0.0030 -0.0040  0.0078
RANGE_HIGH      0.0006  0.0007  0.0011 -0.0016  0.0028
SST            -0.0057 -0.0058  0.0045 -0.0142  0.0033
ELEVATION       0.0001  0.0001  0.0001  0.0000  0.0002
Year2015        0.0808  0.0810  0.0180  0.0450  0.1154
RANGE_LOW:SST  -0.0001 -0.0001  0.0001 -0.0003  0.0001
rho             0.0756  0.0757  0.0151  0.0466  0.1054
sig2eps         0.0054  0.0054  0.0002  0.0051  0.0057
sig2eta         0.0764  0.0739  0.0121  0.0617  0.1073
phi             0.0501  0.0502  0.0090  0.0322  0.0659
------------------------------------------------------------
```

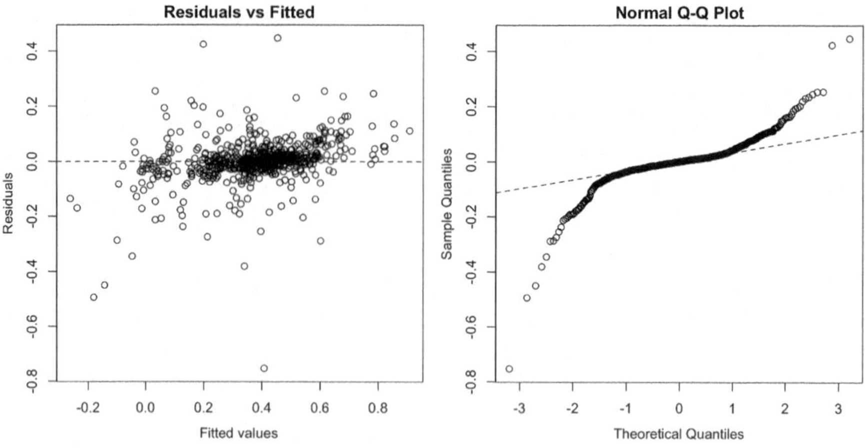

Fig. 7 The residual plot and QQ plot from AR(1) prediction

The dummy variable for year 2015 is significant. After back-transforming, we can say the predicted monthly rainfall for 2015 is exp(0.0808) = 1.084 times greater than the predicted monthly rainfall in other years, holding other predictors fixed. This is consistent with the flooding event in the fall of 2015. Another finding is that elevation might be an explanatory factor to the rainfall since higher elevation relates to higher volumes of precipitation. In addition, a statistically significant and positive ρ indicates that a rainy month might tend to precede another one. On the other hand, the SST has a marginally negative effect on the rainfall prediction but is not significant based on the 95% credible interval.

We also obtain the residuals and the QQ plot in Fig. 7. There is no obvious pattern in the residual plots. However, the residuals show deviations in the tails to some extent from normality based on the QQ plot on the right panel, which indicates a heavy-tailed error distribution and lack of symmetrical pattern (e.g., Samadi et al. 2017).

6 Model Comparison: State-Space Model vs. Gaussian Process

Another framework for spatio-temporal data analysis is the dynamic state-space model. A formulation of the spatio-temporal framework (Stroud et al. 2001) is specified as follows:

$$y_t(\mathbf{s}) = \mathbf{x}_t(\mathbf{s})^T \boldsymbol{\beta}_t + u_t(\mathbf{s}) + \epsilon_t(\mathbf{s}), \quad \epsilon_t(\mathbf{s}) \sim N(0, \tau_t^2)$$

$$\boldsymbol{\beta}_t = \boldsymbol{\beta}_{t-1} + \boldsymbol{\eta}_t, \quad \boldsymbol{\eta}_t \sim N(0, \boldsymbol{\Sigma}_\eta)$$

$$\mu_t(\mathbf{s}) = \mu_{t-1}(\mathbf{s}) + w_t(\mathbf{s}), \quad w_t(\mathbf{s}) \sim GP(\mathbf{0}, C_t(\cdot, \boldsymbol{\theta}_t)).$$

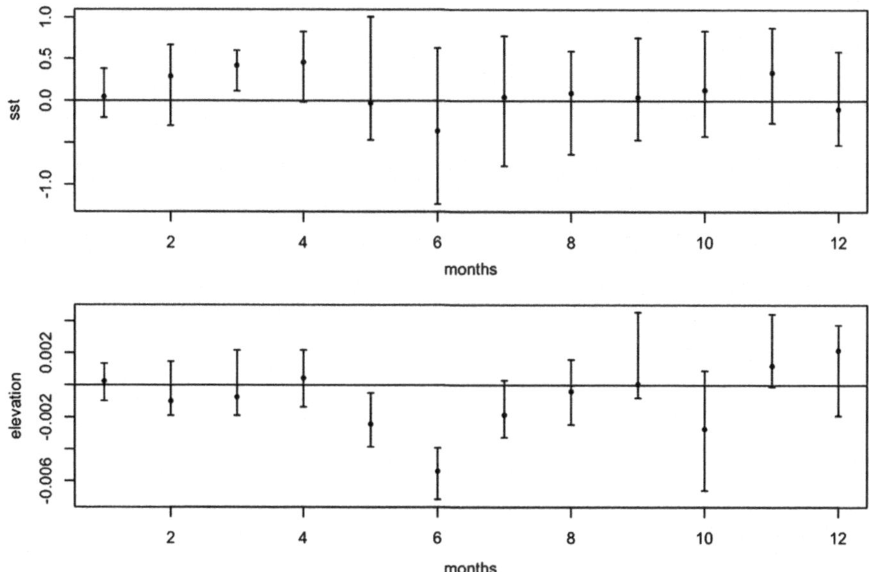

Fig. 8 The 95% confidence interval for β_1 (the SST-related variable) and β_2 (elevation) over 12 months in 2015

Here $\mathbf{x}_t(\mathbf{s})$ is a $p \times 1$ vector of predictors and $\boldsymbol{\beta}_t$ is a $p \times 1$ vector of coefficients. The $GP(\mathbf{0}, C_t(\cdot, \boldsymbol{\theta}_t))$ denotes a spatial Gaussian process with covariance function $C_t(\cdot, \boldsymbol{\theta}_t)$. We further specify $C_t(\mathbf{s}_1, \mathbf{s}_2; \boldsymbol{\theta}_t)) = \sigma_t^2 \rho(\mathbf{s}_1, \mathbf{s}_2; \phi_t)$, where $\boldsymbol{\theta}_t = \{\sigma_t^2, \phi_t\}$ and $\rho(\cdot; \phi)$ is a correlation function with ϕ controlling the correlation decay.

The same response variable and covariates with AR(1) model are used when fitting the state-space model. The R package spBayes (Finley et al. 2007) provides a framework to sample from parameters and posterior. The 95% credible interval for sea surface temperature and elevation are plotted for all 12 months in 2015.

The state-space model allows for a more detailed monthly look of the effect of covariates. For instance, one can conclude that, based on Fig. 8, the SST-based variable effects the rainfall amount in a more significant manner during the first few months of the year. These results strengthen the previous findings of Häkkinen (2000), Mehta et al. (2000), Wang et al. (2006), and Dima and Lohmann (2010), and further support the hypothesis that the variability of North Atlantic SST is coherent with the fluctuations of the rainfall pattern and occurrence. In other words, intense ocean–atmosphere coupling exists in the North Atlantic, particularly during winter. In contrast, elevation is more related to the precipitation in June and October, when heavier rainfall data are observed. This covariate specifies a convective mode that is widely recognized as an important contributor to the probability and type of severe convective rainfall during summer and early fall in the southeast region. The residual plot and the QQ plot for the state-space are shown in Fig. 9. We see the heavy-tailed error pattern is still apparent in this model, based on the QQ plot.

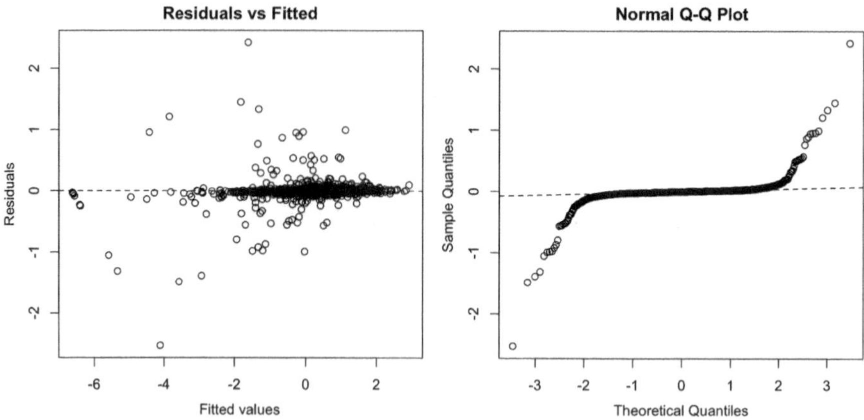

Fig. 9 The residual plot and QQ plot from the state-space model

7 Discussion

We have presented both spatial and spatio-temporal models for rainfall in South Carolina during a period including one of the most destructive storms in state history. Our models have allowed us to determine several covariates that affect the rainfall and to interpret their effects. In particular, the flood year of 2015 was an important indicator of rainfall and elevation also had a positive significant effect on precipitation. There was a significant positive correlation in rainfall measurements over time. Finally, our novel SST index provided some evidence that cooler nearby sea temperatures corresponded to higher rainfall at in land sites although this SST effect was not significant at the 0.05 level based on a 95% credible interval for its effect.

A spatial prediction at a new location and a temporal prediction at a future time point can be obtained based on the posterior predictive distribution for $Z(s_0, t')$, where s_0 denotes a new location and t' is a future time point. Further details regarding these predictions are provided in Cressie and Wikle (2015) for the GP models, and Sahu and Bakar (2012) for the AR models.

A limitation of the study, and a direction for future research, is that the model does not account for the apparent heavy-tailed nature of the errors. Methods involving generalized extreme value distribution (Rodríguez et al. 2016) could possibly be adapted to this model to help handle this heavy-tailed error structure, but such research is still relatively new in the spatio-temporal modeling literature.

References

Anderson, T.W.: An Introduction to Multivariate Statistical Analysis. Wiley, Hoboken (2003)
Bakar, K.S., Sahu, S.K.: spTimer: spatio-temporal Bayesian modeling using R. J. Stat. Softw. **63**(15), 1–32 (2015)

Banerjee, S., Carlin, B.P., Gelfand, A.E.: Hierarchical Modeling and Analysis for Spatial Data. CRC Press, Boca Raton (2014)

Benzécri, J.P.: L'Analyse des Données. Dunod, Paris (1973)

Berne, A., Delrieu, G., Boudevillain, B.: Variability of the spatial structure of intense Mediterranean precipitation. Adv. Water Resour. **32**(7), 1031–1042 (2009)

Bivand, R.S., Pebesma, E.J., Gomez-Rubio, V., Pebesma, E.J.: Applied Spatial Data Analysis with R. Springer, New York (2008)

Ciach, G.J., Krajewski, W.F.: Analysis and modeling of spatial correlation structure in small-scale rainfall in central Oklahoma. Adv. Water Resour. **29**(10), 1450–1463 (2006)

Cressie, N.: Statistics for Spatial Data. Wiley, New York (1993)

Cressie, N., Wikle, C.K.: Statistics for Spatio-Temporal Data. Wiley, New York (2015)

Deidda, R.: Rainfall downscaling in a space-time multifractal framework. Water Resour. Res. **36**(7), 1779–1794 (2000)

Delhomme, J.P.: Kriging in the hydrosciences. Adv. Water Resour. **1**(5), 251–266 (1978)

Delfiner, P., Delhomme, J.P.: Optimum Interpolation by Kriging. Ecole Nationale Supérieure des Mines, Paris (1975)

Diggle, P.J., Tawn, J.A., Moyeed, R.A.: Model-based geostatistics. J. R. Stat. Soc.: Ser. C: Appl. Stat. **47**(3), 299–350 (1998)

Dima, M., Lohmann, G.: Evidence for two distinct modes of large-scale ocean circulation changes over the last century. J. Clim. **23**(1), 5–16 (2010)

Dumitrescu, A., Birsan, M.V., Manea, A.: Spatio-temporal interpolation of sub-daily (6 h) precipitation over Romania for the period 1975–2010. Int. J. Climatol. **36**(3), 1331–1343 (2016)

Ferraris, L., Gabellani, S., Rebora, N., Provenzale, A.: A comparison of stochastic models for spatial rainfall downscaling. Water Resour. Res. **39**(12), 1368 (2003). https://doi.org/10.1029/2003WR002504

Finley, A.O., Banerjee, S., Carlin, B.P.: spBayes: an R package for univariate and multivariate hierarchical point-referenced spatial models. J. Stat. Softw. **19**(4), 1 (2007)

Gelfand, A.E., Diggle, P., Guttorp, P., Fuentes, M.: Handbook of Spatial Statistics. CRC Press, Boca Raton (2010)

Georgakakos, K.P., Kavvas, M.L.: Precipitation analysis, modeling, and prediction in hydrology. Rev. Geophys. **25**(2), 163–178 (1987)

Häkkinen, S.: Decadal air-sea interaction in the North Atlantic based on observations and modeling results. J. Clim. **13**(6), 1195–1219 (2000)

Hastings, W.K.: Monte Carlo sampling methods using Markov chains and their applications. Biometrika **57**(1), 97–109 (1970)

Hughes, J., Haran, M.: Dimension reduction and alleviation of confounding for spatial generalized linear mixed models. J. R. Stat. Soc. Ser. B Stat Methodol. **75**(1), 139–159 (2013)

Isaaks, H.E., Srivastava, R.M.: Applied Geostatistics. Oxford University Press, New York (1989)

Kumar, P., Foufoula-Georgiou, E.: Characterizing multiscale variability of zero intermittency in spatial rainfall. J. Appl. Meteorol. **33**(12), 1516–1525 (1994)

Ly, S., Charles, C., Degre, A.: Geostatistical interpolation of daily rainfall at catchment scale: the use of several variogram models in the Ourthe and Ambleve catchments, Belgium. Hydrol. Earth Syst. Sci. **15**(7), 2259–2274 (2011)

Matheron, G.: Principles of geostatistics. Econ. Geol. **58**(8), 1246–1266 (1963)

Mehta, V., Suarez, M., Manganello, J.V., Delworth, T.D.: Oceanic influence on the North Atlantic oscillation and associated northern hemisphere climate variations: 1959–1993. Geophys. Res. Lett. **27**(1), 121–124 (2000)

Metropolis, N., Rosenbluth, A.W., Rosenbluth, M.N., Teller, A.H., Teller, E.: Equation of state calculations by fast computing machines. J. Chem. Phys. **21**(6), 1087–1092 (1953)

National Oceanic and Atmosphere Administration, U.S. Department of Commerce: Service assessment: the historic South Carolina floods of October 1–5, 2015. www.weather.gov/media/publications/assessments/SCFlooding_072216_Signed_Final.pdf (2015). Accessed 4 Dec 2017

Rodríguez, S., Huerta, G., Reyes, H.: A study of trends for Mexico city ozone extremes: 2001–2014. Atmósfera **29**(2), 107–120 (2016)

Sahu, S.K., Bakar, K.S.: Hierarchical Bayesian autoregressive models for large space–time data with applications to ozone concentration modeling. Appl. Stoch. Model. Bus. Ind. **28**(5), 395–415 (2012)

Samadi, S., Tufford, D., Carbone, G.: Estimating hydrologic model uncertainty in the presence of complex residual error structures. Stoch. Environ. Res. Risk Assess. **32**(5), 1259–1281 (2018)

Sharon, D.: Spatial analysis of rainfall data from dense networks. Hydrol. Sci. J. **17**(3), 291–300 (1972)

Stroud, J.R., Müller, P., Sansó, B.: Dynamic models for spatio-temporal data. J. R. Stat. Soc. Ser. B Stat. Methodol. **63**(4), 673–689 (2001)

Tabios III, Q.G., Salas, J.D.: A comparative analysis of techniques for spatial interpolation of precipitation. Water Resour. Bull. **21**(3), 365–380 (1985)

Thiessen, A.H.: Precipitation averages for large areas. Mon. Weather Rev. **39**(7), 1082–1084 (1911)

Troutman, B.M.: Runoff prediction errors and bias in parameter estimation induced by spatial variability of precipitation. Water Resour. Res. **19**(3), 791–810 (1983)

Wang, C., Enfield, D.B., Lee, S.K., Landsea, C.W.: Influences of the Atlantic warm pool on western hemisphere summer rainfall and Atlantic hurricanes. J. Clim. **19**(12), 3011–3028 (2006)

A Sparse Areal Mixed Model for Multivariate Outcomes, with an Application to Zero-Inflated Census Data

Donald Musgrove, Derek S. Young, John Hughes, and Lynn E. Eberly

1 Introduction

The Committee on National Statistics assembled the Panel to Review the 2010 Census to suggest general priorities for research in preparation for the 2020 Census. In their first interim report (Cook et al. 2011) the Panel laid out three recommendations, the first of which highlighted "four priority topic areas, in order to achieve a lower cost and high-quality 2020 Census." A theme across these priority areas was the effective use of Census Bureau databases (e.g., geographic databases and databases built with administrative records) to achieve operational objectives. In addition to implementing recommendations from the Panel to Review the 2010 Census, the Census Bureau is placing increasing emphasis on accurate model-based predictions as a way to more generally conduct efficient and cost-effective surveys (U.S. Census Bureau 2015).

One of the Census Bureau's most prominent databases is the Master Address File (MAF), which is a continually updated inventory of all known living quarters in the USA and its island territories. The MAF is used as a sampling frame for various Census Bureau surveys, including the decennial Census. The MAF comprises

D. Musgrove
Medtronic, Minneapolis, MN, USA

D. S. Young (✉)
Department of Statistics, University of Kentucky, Lexington, KY, USA
e-mail: derek.young@uky.edu

J. Hughes
Department of Biostatistics and Informatics, University of Colorado, Denver, CO, USA

L. E. Eberly
Division of Biostatistics, University of Minnesota, Minneapolis, MN, USA

© Springer Nature Switzerland AG 2019

N. Diawara (ed.), *Modern Statistical Methods for Spatial and Multivariate Data*,
STEAM-H: Science, Technology, Engineering, Agriculture, Mathematics & Health,
https://doi.org/10.1007/978-3-030-11431-2_3

approximately 194 million records, and includes geographic information, residential status codes, and addresses. As discussed in Young et al. (2017), there is interest in building census-block-level models to (1) help understand coverage errors resulting from a particular MAF extract, which is what defines a survey sampling frame, and (2) identify blocks with a large number of expected housing unit "adds" and "deletes," which could aid targeted address canvassing operations.

Our analysis uses data from the 2010 Census. We are interested in revealing causes of address deletes, which are defined by the Census Bureau as addresses that were deleted from the base count because they did not correspond to valid housing units (a house, an apartment, a mobile home, a group of rooms, or a single room that is occupied, or intended for occupancy, as a separate living area). The data are zero-inflated since the aggregated number of deletes has over 90% zeros across the entire USA. Some zeros arise from those areas in which changes in mailing addresses are unlikely. Another source of zeros is areas prone to redevelopment, in which case there is a high "risk" of future address changes. We include various geographic, demographic, and operational variables as predictors in our model. This work could lead to a more efficient and cost-effective decennial Census.

The source of the data we analyze is a publicly available Census Bureau dataset called the Planning Database (PDB) (http://goo.gl/LlcwY7). These data include variables and counts from the 2010 Census and the 2009–13 American Community Survey (ACS). The data are aggregated at the block-group level rather than the block level. A census block is the smallest geographic unit used by the Census Bureau. Blocks are typically bounded by streets or creeks/rivers. Within a city, a block corresponds to a city block. In a rural area, blocks may be large and irregularly shaped and bounded by features including roads, streams, or transmission lines. Census blocks are not delineated based on population. A block group, on the other hand, comprises multiple blocks and contains between 600 and 3000 people. The PDB comprises approximately 220,000 block groups and 300 variables.

Young et al. (2017) developed zero-inflated models to reveal predictors of housing unit adds and deletes, but their data were from a particular MAF extract. Their work highlighted the potential success of using model-based predictions in the decision-making process of costly Census Bureau operations. However, the models of Young et al. were developed based on data collected just before and just after the Census Bureau's 2010 address canvassing operation. Thus, their results illustrated what could have been done for 2010 operations using model-based strategies. Regardless, their work served as a *proof-of-concept* that model-based strategies should be investigated using other federal databases. For example, the PDB is an excellent candidate because it includes more geographic and demographic variables than are available from a traditional MAF extract. We also note that the models developed in Young et al. (2017) were non-spatial. Failure to account for consequential spatial dependence can lead to erroneous inference, e.g., confidence intervals for regression coefficients may be too narrow. Thus, we develop a sparse areal mixed model for multivariate outcomes to reflect the type of features in the data discussed above.

Multivariate areal (i.e., spatially aggregated) outcomes are common in a number of fields. Disease mapping provides what is perhaps the canonical example: a number of disease outcomes (e.g., incidence, prevalence, and rate) that share a collection of spatially varying risk factors. Such outcomes are likely to exhibit dependence both within and across areal units, in which case applying a multivariate model instead of multiple univariate models can yield improved inference. For our Census data analysis, we assess a collection of spatially varying covariates when modeling the number of address deletes. We accomplish this goal by developing a new mixed model for multivariate areal data.

Traditional mixed models for multivariate areal outcomes face two formidable challenges: (1) spatial confounding (Clayton et al. 1993; Reich et al. 2006), which can lead to erroneous regression inference, and (2) an immense computational burden. Our approach addresses both of these challenges by extending the sparse areal mixed model (SAMM) of Hughes and Haran (2013). The model is further specialized to handle zero-inflated data.

The remainder of this chapter is organized as follows. In Sect. 2 we develop our model, the multivariate sparse areal mixed model (MSAMM). In Sect. 3 we discuss the hyperpriors used and computational details for the MSAMM. In Sect. 4 we specialize the MSAMM to handle zero-inflated data and apply the resulting model to zero-inflated Census data for the state of Iowa. In Sect. 5 we conclude with a summary and a sketch of future work. We provide derivations, computational details, and extended simulation results in an appendix.

2 Our Sparse Mixed Model for Multivariate Areal Data

In this section we develop our multivariate sparse areal mixed model (MSAMM). Our approach is similar to the approach of Bradley et al. (2015) in that we, too, employ the orthogonal, multiresolutional spatial basis described by Hughes and Haran (2013) (see also Griffith (2003) and Tiefelsdorf and Griffith (2007)). This basis, known as the Moran (1950) basis, is appealing from a modeling point of view and also permits efficient computing.

2.1 Review of Univariate CAR Models

To motivate our development of the MSAMM, we begin by reviewing conditional autoregressive (CAR) models for univariate areal data. The transformed conditional mean vector for these models is given by

$$g(\mu) = g\left\{\mathbb{E}\left(y \mid \beta, \, \phi\right)\right\} = \mathbf{X}\beta + \phi, \tag{1}$$

where g is a link function; $y = (y_1, \ldots, y_n)'$ are the outcomes, the ith of which is associated with the ith areal unit; $\mathbf{X}$ is an $n \times p$ design matrix; β is a p-vector of regression coefficients; and $\boldsymbol{\phi} = (\phi_1, \ldots, \phi_n)'$ are spatially dependent random effects. Note that the outcomes are assumed to be independent conditional on the random effects. Marginally, however, the outcomes are spatially dependent because the random effects are spatially dependent.

The various CAR models are distinguished by the distribution assigned to $\boldsymbol{\phi}$. The most popular specification for $\boldsymbol{\phi}$ is the intrinsic CAR model (Besag and Kooperberg 1995):

$$p(\boldsymbol{\phi} \mid \tau) \propto \exp\left(-\frac{\tau}{2}\boldsymbol{\phi}'\mathbf{Q}\boldsymbol{\phi}\right),$$

where $p(\cdot)$ denotes a prior distribution, τ is a smoothing parameter, and the precision matrix $\mathbf{Q}$ is equal to $\mathbf{D}-\mathbf{A}$, where $\mathbf{D}$ is the diagonal matrix with the degrees (i.e., number of neighbors) of the areal units on its diagonal and $\mathbf{A}$ is the adjacency matrix for the underlying graph G, i.e., $\mathbf{A}$ is the binary matrix that encodes the adjacency structure among the areal units. Since $\mathbf{Q}$ is singular, the intrinsic CAR is improper. A proper alternative has precision matrix $\mathbf{Q}(\rho) = \mathbf{D} - \rho\mathbf{A}$, where ρ is constrained to the interval $[0, 1)$. The parameter ρ can be considered a range parameter, but its effect on the marginal dependence structure is complex and often pathological (Wall 2004; Assunção and Krainski 2009). Additional proper CAR specifications exist. For instance, the prior proposed by Leroux et al. (2000) offers an alternative specification of the spatial random effects resulting in a proper prior distribution.

Traditional CAR models present serious challenges. First, there is often multicollinearity between the spatial random effects and the fixed-effects predictors. This characteristic, which is known as spatial confounding (Clayton et al. 1993), often leads to (1) biased estimation of regression coefficients and (2) substantial variance inflation that may make important covariates appear insignificant. Additionally, CAR random effects permit patterns of spatial repulsion, which we do not expect to observe in the types of data to which these models are usually applied.

Second, computation for CAR models can be extremely burdensome due to (1) the high dimensionality of $\boldsymbol{\phi}$ and (2) the nature of $\boldsymbol{\phi}$'s posterior distribution. It is well known that a univariate Metropolis–Hastings algorithm for sampling from the posterior distribution of $\boldsymbol{\phi}$ leads to a slow mixing Markov chain because the components of $\boldsymbol{\phi}$ exhibit strong *a posteriori* dependence. This has led to a number of approaches for updating $\boldsymbol{\phi}$ in a block(s). Constructing proposals for these updates is challenging, and the faster mixing comes at the cost of increased running time per iteration, see, for instance, Knorr-Held and Rue (2002), Haran et al. (2003), and Haran and Tierney (2012).

2.2 Review of Hughes and Haran's SAMM

To alleviate spatial confounding, eliminate patterns of spatial repulsion, and greatly reduce computing time and storage requirements, Hughes and Haran (2013) introduced their sparse areal mixed model (SAMM). In signal processing, statistics, and related fields, it is not uncommon to use the term "sparse" to refer to representation of a signal in terms of a small number of generating elements drawn from an appropriately chosen domain (Donoho and Elad 2003). We use the term "sparse" in precisely this sense, since our model accomplishes spatial smoothing by using $q \ll n$ Moran basis vectors (as opposed to traditional CAR models, which have approximately n spatial random effects). The SAMM can be developed as follows.

Reich et al. (2006) showed that the traditional CAR models are spatially confounded in the sense that the random effects can "pollute" the regression manifold $C(\mathbf{X})$, which can lead to a biased and variance-inflated posterior for $\boldsymbol{\beta}$. To see this, first let $\mathbf{P}$ be the orthogonal projection onto $C(\mathbf{X})$, so that $\mathbf{I}_n - \mathbf{P}$ is the orthogonal projection onto $C(\mathbf{X})^{\perp}$. Now eigendecompose $\mathbf{P}$ and $\mathbf{I}_n - \mathbf{P}$ to obtain orthogonal bases $\mathbf{K}_{n \times p}$ and $\mathbf{L}_{n \times (n-p)}$ for $C(\mathbf{X})$ and $C(\mathbf{X})^{\perp}$, respectively. Then (1) can be rewritten as

$$g(\boldsymbol{\mu}) = \mathbf{X}\boldsymbol{\beta} + \mathbf{K}\boldsymbol{\gamma} + \mathbf{L}\boldsymbol{\delta},$$

where $\boldsymbol{\gamma}_{p \times 1}$ and $\boldsymbol{\delta}_{(n-p) \times 1}$ are random coefficients. This form shows that $\mathbf{K}$ is the source of the confounding, for $\mathbf{K}$ and $\mathbf{X}$ have the same column space.

Since the columns of $\mathbf{K}$ are merely synthetic predictors (i.e., they have no scientific meaning), Reich et al. (2006) recommend removing them from the model. The resulting model (henceforth the RHZ model) has

$$g(\boldsymbol{\mu}) = \mathbf{X}\boldsymbol{\beta} + \mathbf{L}\boldsymbol{\delta},$$

so that spatial smoothing is restricted to the orthogonal complement of $C(\mathbf{X})$. In a subsequent paper, Hodges and Reich (2010) referred to this technique as restricted spatial regression (RSR).

RSR is not only an effective remedy for confounding but also speeds computing. Because the columns of $\mathbf{L}$ are orthogonal, the RHZ model's random effects are approximately *a posteriori* uncorrelated. This yields a fast-mixing Markov chain, and the cost per iteration is reduced because a simple spherical normal proposal is sufficient for updating the random effects. But fitting the RHZ model to large areal datasets is still quite burdensome computationally because the random effects remain high dimensional.

By taking full advantage of the underlying graph G, Hughes and Haran (2013) were able to greatly reduce the number of random effects while also improving regression inference. Hughes and Haran (2013) begin by defining the so-called Moran operator for $\mathbf{X}$ with respect to G: $(\mathbf{I}_n - \mathbf{P})\mathbf{A}(\mathbf{I}_n - \mathbf{P})$. This operator appears in the numerator of a generalized form of Moran's I, a popular nonparametric measure of spatial dependence for areal data (Moran 1950):

$$I_{\mathbf{X}}(\mathbf{A}) = \frac{n}{\mathbf{1}'\mathbf{A}\mathbf{1}} \frac{\mathbf{y}'(\mathbf{I}_n - \mathbf{P})\mathbf{A}(\mathbf{I}_n - \mathbf{P})\mathbf{y}}{\mathbf{y}'(\mathbf{I}_n - \mathbf{P})\mathbf{y}}.$$

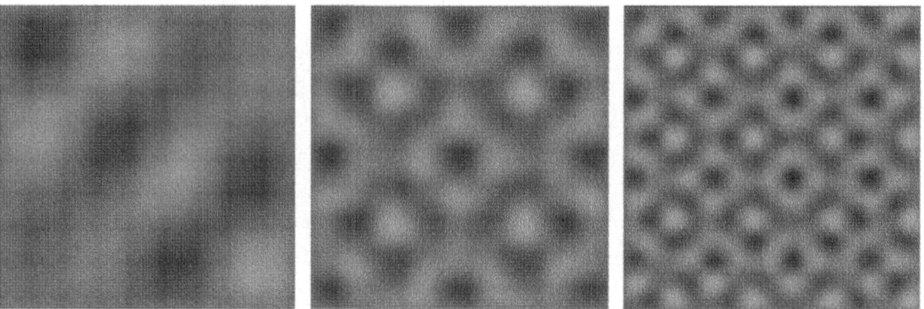

Fig. 1 Three Moran basis vectors, exhibiting spatial patterns of increasingly finer scale

Boots and Tiefelsdorf (2000) showed that (1) the (standardized) spectrum of a Moran operator comprises the possible values for the corresponding $I_\mathbf{X}(\mathbf{A})$, and (2) the eigenvectors comprise all possible mutually distinct patterns of clustering residual to $C(\mathbf{X})$ and accounting for G. The positive (negative) eigenvalues correspond to varying degrees of positive (negative) spatial dependence, and the eigenvectors associated with a given eigenvalue (ω_i, say) are the patterns of spatial clustering that data exhibit when the dependence among them is of degree ω_i.

In other words, the eigenvectors of the Moran operator form a multiresolutional spatial basis for $C(\mathbf{X})^\perp$ that exhausts all possible patterns that can arise on G. Since we do not expect to observe repulsion in the phenomena to which these models are usually applied, we can use the spectrum of the operator to discard all repulsive patterns, retaining only attractive patterns for our analysis. By retaining only eigenvectors that exhibit positive spatial dependence, we can usually reduce the model dimension by at least half a priori. Hughes and Haran (2013) showed that a much greater reduction is often possible in practice, with 50–100 eigenvectors being sufficient in many cases. Three example Moran vectors are shown in Fig. 1.

Let $\mathbf{M}_{n \times q}$ contain the first $q \ll n$ eigenvectors of the Moran operator. Then the SAMM has first stage

$$g(\boldsymbol{\mu}) = \mathbf{X}\boldsymbol{\beta} + \mathbf{M}\boldsymbol{\delta}_s,$$

where $\boldsymbol{\delta}_s$ ("s" for "sparse") is a q-vector of random coefficients that are assumed to be jointly Gaussian:

$$\boldsymbol{\delta}_s \sim \mathcal{N}\{\mathbf{0}, (\tau \mathbf{Q}_s)^{-1}\}, \tag{2}$$

where $\mathbf{Q}_s = \mathbf{M}'\mathbf{Q}\mathbf{M}$. This implies $p + q + 1$ unknowns, compared to $p + n + 1$ for the traditional model and $p + (n - p) + 1 = n + 1$ for the RHZ model. This dramatic reduction in dimension speeds computation considerably, allowing even the largest areal datasets to be analyzed quickly (in minutes or hours rather than days or weeks).

Since $\boldsymbol{\delta}_s$ are regression coefficients, one may be tempted to assign $\boldsymbol{\delta}_s$ a spherical Gaussian prior instead of the abovementioned prior. This would be a mistake, however, for (2) is not arbitrary (see Reich et al. (2006) and/or Hughes and Haran (2013) for derivations) but is, in fact, very well suited to the task at hand. Specifically, two characteristics of (2) discourage overfitting even when q is too large for the dataset being analyzed. First, the prior variances are commensurate with the spatial scales of the predictors in $\mathbf{M}$. This shrinks toward zero the coefficients corresponding to predictors that exhibit small-scale spatial variation. Additionally, the correlation structure of (2) effectively reduces the degrees of freedom in the smoothing component of the model.

2.3 Review of Multivariate CAR Models

A number of multivariate CAR (MCAR) models have been developed (Carlin and Banerjee 2003; Gelfand and Vounatsou 2003; Jin et al. 2005; Martinez-Beneito 2013). These models have the same drawbacks as their univariate counterparts, but of course entail even more burdensome computation. Thus it is desirable to develop a SAMM for multivariate outcomes. We begin by reviewing the MCAR model that is the multivariate analog of the traditional univariate CAR model described above.

Suppose we observe multiple outcomes at each areal unit and that each outcome has its own regression component and collection of spatial effects. Specifically, for $j \in \{1, \ldots, J\}$ we have outcomes $\boldsymbol{y}_j = (y_{1j}, \ldots, y_{nj})'$, design matrix $\mathbf{X}_j$, regression coefficients $\boldsymbol{\beta}_j$, and spatial effects $\boldsymbol{\phi}_j = (\phi_{1j}, \ldots, \phi_{nj})'$. Then the transformed conditional mean vectors are given by

$$g_j(\boldsymbol{\mu}_j) = \mathbf{X}_j \boldsymbol{\beta}_j + \boldsymbol{\phi}_j.$$

Now collect the $\boldsymbol{\phi}_j$ to form $\boldsymbol{\Phi} = (\boldsymbol{\phi}_1', \ldots, \boldsymbol{\phi}_J')'$, and put

$$p(\boldsymbol{\Phi} \mid \boldsymbol{\Sigma}) \propto \exp\left\{-\frac{1}{2}\boldsymbol{\Phi}'\left(\boldsymbol{\Sigma}^{-1} \otimes \mathbf{Q}\right)\boldsymbol{\Phi}\right\},$$

where $\boldsymbol{\Sigma}$ is a $J \times J$ covariance matrix consisting of $J(J+1)/2$ unknown variance and covariance parameters, and $\otimes$ denotes the Kronecker product. The jth diagonal entry of $\boldsymbol{\Sigma}$ is proportional to the variance of the spatial effects corresponding to the jth outcome. The jj' off-diagonal entry is proportional to the covariance between the jth and j'th spatial effects within an areal unit.

Should we require a different precision matrix for each $\boldsymbol{\phi}_j$, the prior on $\boldsymbol{\Phi}$ can be generalized as

$$p(\boldsymbol{\Phi} \mid \boldsymbol{\Sigma}) \propto \exp\left\{-\frac{1}{2}\boldsymbol{\Phi}'\mathrm{bdiag}(\mathbf{R}_1, \ldots, \mathbf{R}_J)'(\boldsymbol{\Sigma}^{-1} \otimes \mathbf{I}_n)\mathrm{bdiag}(\mathbf{R}_1, \ldots, \mathbf{R}_J)\boldsymbol{\Phi}\right\},$$

where $\mathrm{bdiag}(\cdot)$ denotes a block diagonal matrix and $\mathbf{R}_j$ $(j = 1, \ldots, J)$ is such that $\mathbf{R}_j'\mathbf{R}_j = \mathbf{Q}_j$.

2.4 The Multivariate SAMM (MSAMM)

Recently, Bradley et al. (2015) introduced the Moran's I (MI) prior, which is a multivariate spatiotemporal model based on the SAMM. We introduce a multivariate model that uses a similar prior but is strictly for spatial data. We call our model the multivariate SAMM (MSAMM). The MSAMM serves as the foundation for the zero-inflated count model that we focus on below in Sect. 4.

Construction of the MSAMM is of course analogous to construction of the SAMM. For $j \in \{1, \ldots, J\}$, let $\mathbf{P}_j = \mathbf{X}_j(\mathbf{X}'_j\mathbf{X}_j)^{-1}\mathbf{X}'_j$, and let $\mathbf{M}_j$ be a matrix, the columns of which are the first q eigenvectors of $(\mathbf{I}_n - \mathbf{P}_j)\mathbf{A}(\mathbf{I}_n - \mathbf{P}_j)$. Denote the prior precision matrix as $\mathbf{Q}_{sj} = \mathbf{M}'_j\mathbf{Q}\mathbf{M}_j$, and let $\mathbf{R}_{sj}$ be the upper Cholesky triangle of $\mathbf{Q}_{sj}$ so that $\mathbf{R}'_{sj}\mathbf{R}_{sj} = \mathbf{Q}_{sj}$. Then the MSAMM can be specified as

$$g_j(\boldsymbol{\mu}_j) = g_j\left\{\mathbb{E}\left(\mathbf{y}_j \mid \boldsymbol{\beta}_j, \boldsymbol{\delta}_{sj}\right)\right\} = \mathbf{X}_j\boldsymbol{\beta}_j + \mathbf{M}_j\boldsymbol{\delta}_{sj}$$

$$p(\boldsymbol{\Delta} \mid \boldsymbol{\Sigma}) \propto \exp\left\{-\frac{1}{2}\boldsymbol{\Delta}'\mathbf{R}'\left(\boldsymbol{\Sigma}^{-1} \otimes \mathbf{I}_q\right)\mathbf{R}\boldsymbol{\Delta}\right\},$$

where $\boldsymbol{\Delta} = (\boldsymbol{\delta}'_{s1}, \ldots, \boldsymbol{\delta}'_{sJ})'$ and $\mathbf{R} = \mathrm{bdiag}\,(\mathbf{R}_{s1}, \ldots, \mathbf{R}_{sJ})$. Once again $\boldsymbol{\Sigma}$ is a $J \times J$ covariance matrix consisting of $J(J+1)/2$ unknown parameters. Thus the MSAMM has $J(p+q+(J+1)/2)$ unknowns, while the MCAR has $J(p+n+(J+1)/2)$. This is a considerable reduction so long as $q \ll n$. Moreover, the MSAMM, like the SAMM, alleviates spatial confounding; permits simple, fast updating of the spatial random effects; and yields a fast-mixing Markov chain.

In our application and simulation study, we used the same design matrix for all dimensions, i.e., we used $\mathbf{X}_j = \mathbf{X}$ for all j, which implies a single prior precision matrix $\mathbf{Q}_s$. In this case, the above specification simplifies to

$$g_j(\boldsymbol{\mu}_j) = g_j\left\{\mathbb{E}\left(\mathbf{y}_j \mid \boldsymbol{\beta}_j, \boldsymbol{\delta}_{sj}\right)\right\} = \mathbf{X}\boldsymbol{\beta}_j + \mathbf{M}\boldsymbol{\delta}_{sj} \tag{3}$$

$$p(\boldsymbol{\Delta} \mid \boldsymbol{\Sigma}) \propto \exp\left\{-\frac{1}{2}\boldsymbol{\Delta}'\left(\boldsymbol{\Sigma}^{-1} \otimes \mathbf{Q}_s\right)\boldsymbol{\Delta}\right\}.$$

In either case the precision matrix is invertible, and so the prior distribution is proper.

3 Hyperpriors and Computation for the MSAMM

Although using a truncated Moran basis dramatically reduces the time required to draw samples from the posterior, and the space required to store those samples, this approach does incur the substantial up-front burden of computing and eigendecomposing $(\mathbf{I}_n - \mathbf{P}_j)\mathbf{A}(\mathbf{I}_n - \mathbf{P}_j)$. The efficiency of the former can be increased by storing $\mathbf{A}$ in a sparse format and parallelizing the matrix multiplications. And we can more efficiently obtain the desired basis vectors by computing only the first q eigenvectors of $(\mathbf{I}_n - \mathbf{P}_j)\mathbf{A}(\mathbf{I}_n - \mathbf{P}_j)$ instead of doing the full eigendecomposition. This can be done using the Spectra library (Qiu 2017), for example.

We use independent Gaussian priors for the regression coefficients: $\boldsymbol{\beta}_j \overset{\text{ind}}{\sim} \mathcal{N}(\mathbf{0}, 10,000\,\mathbf{I}_p)$. There are several appealing choices for the prior distribution of $\boldsymbol{\Sigma}$. We follow Huang and Wand (2013) and use the hierarchical half-t prior distribution. The hierarchical half-t prior relies on an inverse-Wishart distribution such that the diagonal elements of the scale matrix are given inverse-gamma prior distributions. The prior scale of the inverse-Wishart is $2\nu\mathbf{V}$, which implies $\nu + J - 1$ degrees of freedom. We let $\mathbf{V} = \mathrm{diag}\,(1/v_1, \ldots, 1/v_J)$, where v_j has an inverse-gamma prior with shape $1/2$ and scale $1/\zeta_j^2$. The hyperparameters are given values of $\nu = 2$ and $\zeta_j = 10^5$ $(j = 1, \ldots, J)$. This approach yields a conjugate prior distribution. Alternatives include the inverse-Wishart prior, the covariance matrix separation strategy (Barnard et al. 2000), and the LKJ prior (Lewandowski et al. 2009).

The full conditional distribution of the spatial effects $\boldsymbol{\Delta}$ does not have a closed form, and so we use a Metropolis update to draw samples from the posterior. When a common design matrix is used for all dimensions, we can speed computation by reparameterizing the spatial effects. Suppose we have $\boldsymbol{\Psi} \sim \mathcal{N}(\mathbf{0}, \boldsymbol{\Sigma} \otimes \mathbf{I}_q)$. Let $\mathbf{R}_s$ be the upper Cholesky triangle of $\mathbf{Q}_s$, and let $\mathbf{W}_s = \mathbf{R}_s^{-1}$ so that $\mathbf{W}_s\mathbf{W}_s' = \mathbf{Q}_s^{-1}$. Then it is easy to see that $(\mathbf{I}_J \otimes \mathbf{W}_s)\boldsymbol{\Psi}$ and $\boldsymbol{\Delta}$ have the same distribution. We replace $\mathbf{M}$ with $\mathbf{M}_s = \mathbf{M}\mathbf{W}_s$ and work with $\boldsymbol{\Psi}$ instead of $\boldsymbol{\Delta}$. We provide details in an appendix. We also show that a similar reparameterization holds when multiple design matrices are required.

For our simulation study and application to the Census data, we used fixed-width analysis (Flegal et al. 2008), in which samples are drawn until all Monte Carlo standard errors are smaller than some pre-selected threshold. We used the `batchmeans` package (Haran and Hughes 2016) for R (Ihaka and Gentleman 1996) to compute Monte Carlo standard errors. Two hundred thousand samples were sufficient to ensure that all Monte Carlo standard errors were less than 0.05. The MCMC estimates stabilized after approximately 100,000 iterations. The time required to fit the MSAMM was approximately equal to the time required to fit independent SAMMs.

Our software was written in R and C++ (Stroustrup 2013). Our use of C++ was aided greatly through the use of the `Rcpp` package (Eddelbuettel and Francois 2011). Most of the numerical linear algebra was carried out using the Armadillo C++ library (Sanderson 2010), which we accessed by way of the `RcppArmadillo` package (Eddelbuettel and Sanderson 2014).

4 An MSAMM for Zero-Inflated Data

In this section we specialize the MSAMM to handle zero-inflated count data. Zero-inflated count models are inherently two-component mixture models, where one component is a point mass at 0 and the other is a discrete distribution. In recent years, many novel finite mixture models have been developed to incorporate spatial dependencies. Alfó et al. (2009) used finite mixture models to analyze multiple, spatially correlated, counts, where the dependence among outcomes is modeled using a set of correlated random effects. Green and Richardson (2002) developed a class of hidden Markov models in the spatial domain to analyze spatial heterogeneity of count data on a rare phenomenon. Neelon et al. (2015) developed a broad class of Bayesian two-part models for the spatial analysis of semicontinuous data. Torabi (2016) proposed a hierarchical multivariate mixture generalized linear model to simultaneously analyze spatial normal and non-normal outcomes. Zero-inflated count models are often applied in non-spatial settings, e.g., in manufacturing (Lambert 1992), where defective materials are rare and the number of defects is assumed to follow a Poisson distribution, and in the hunger-for-bonus phenomenon that occurs in risk assessment for filed insurance claims (Boucher et al. 2009). Spatial zero-inflated count models have been applied to various types of data, including animal sightings (Agarwal et al. 2002; Ver Hoef and Jansen 2007; Recta et al. 2012), plant distribution (Rathbun and Fei 2006), tornado reports (Wikle and Anderson 2003), and emergency room visits (Neelon et al. 2013).

4.1 Models for Zero-Inflated Counts

Two common approaches to modeling zero-inflated counts are the hurdle model and the zero-inflated-Poisson (ZIP) model (Lambert 1992). For a hurdle model, the outcome is 0 with probability $1 - \pi$, and with probability π the outcome arose from a zero-truncated Poisson (ZTP) distribution (Cohen 1960; Singh 1978). Formally, the hurdle model is of the form

$$\mathbb{P}(y = 0) = 1 - \pi$$

$$\mathbb{P}(y = k) = \pi \frac{\exp(-\lambda)}{1 - \exp(-\lambda)} \frac{\lambda^k}{k!} \qquad (k \in \mathbb{N} : k \geq 1).$$

The ZIP model, on the other hand, is given by

$$\mathbb{P}(y = 0) = (1 - \pi) + \pi \exp(-\lambda)$$

$$\mathbb{P}(y = k) = \pi \frac{\lambda^k}{k!} \exp(-\lambda) \qquad (k \in \mathbb{N} : k \geq 1).$$

Let us compare and contrast the hurdle and ZIP models informally. Each model can be viewed as comprising a binary process (the incidence process) and a counting process (the prevalence process). For the hurdle model there is only one source of zeros, namely the binary process. If the binary outcome is 0, no count is observed. If the binary outcome is 1, a nonzero count is observed. The ZIP model differs in that it posits two sources of zeros. If the binary outcome is 0, no count is observed.

If the binary outcome is 1, a (possibly 0) count is observed. This difference has a practical effect: the hurdle model can accommodate both zero-inflation and zero-deflation while the ZIP model can accommodate only zero-inflation (Neelon et al. 2013).

It may also be the case that one or the other of the models makes more sense given the phenomenon of interest. Consider emergency room visits, for example, as in Neelon et al. (2013). A hurdle model is appropriate here; the binary outcome registers whether a given subject visits an emergency room while the count outcome records the number of visits. As a second example, consider animal sightings (Agarwal et al. 2002; Recta et al. 2012). A type of animal may not be sighted in a given region even though the animal is likely present in the region. A ZIP model is appropriate in this scenario. We adopt the hurdle model for the remainder of the paper.

4.2   A Spatial Poisson Hurdle Model Based on the MSAMM

In the context of our MSAMM, an appropriate hurdle model can be specified as follows. Suppose we have n areal units, in which case our outcomes are $y_1, \ldots, y_n$. Although the outcomes are univariate, we employ a bivariate MSAMM for the mean structure, i.e., for the pairs of incidence probabilities and prevalence rates $(\pi_i, \lambda_i)'$ $(i = 1, \ldots, n)$. For the sake of simplicity, suppose that the same design matrix, $\mathbf{X}$, is appropriate for both the incidence process and the prevalence process. This implies an MSAMM of the form given in (3) (as we showed in Sect. 2.3, this model can easily be extended to accommodate different design matrices). Thus the linear predictors for the ith areal unit can be specified as

$$\eta_{i1} = x_i' \boldsymbol{\beta}_1 + m_i' \boldsymbol{\delta}_{s1} \quad \text{(incidence process)}$$

$$\eta_{i2} = x_i' \boldsymbol{\beta}_2 + m_i' \boldsymbol{\delta}_{s2} \quad \text{(prevalence process)}.$$

Using the logit and log link functions, respectively, gives

$$\pi_i = \frac{\exp(\eta_{i1})}{1 + \exp(\eta_{i1})}$$

$$\lambda_i = \exp(\eta_{i2}),$$

where π_i is the probability of incidence for the ith areal unit, and λ_i is the ZTP rate for the ith areal unit. The within-unit covariance matrix $\boldsymbol{\Sigma}$ is of course 2×2 for this model. Clearly, this model accommodates (1) spatial dependence among areal units, and (2) dependence between the incidence process and the prevalence process. For the case of a positive off-diagonal value in $\boldsymbol{\Sigma}$, the latter source of dependence implies that a higher probability of incidence is associated with a higher prevalence rate. As we will see in the next section, our MSAMM hurdle model's ability to accommodate consequential dependence between π_i and λ_i permits improved inference and fit.

4.3   Simulated Application of the MSAMM Hurdle Model

To assess the performance of our areal hurdle model, we carried out a simulation study. We simulated data for the 2600 census block groups of the US state of Iowa (Fig. 2). We included an intercept term and, as a covariate, the percentage of housing

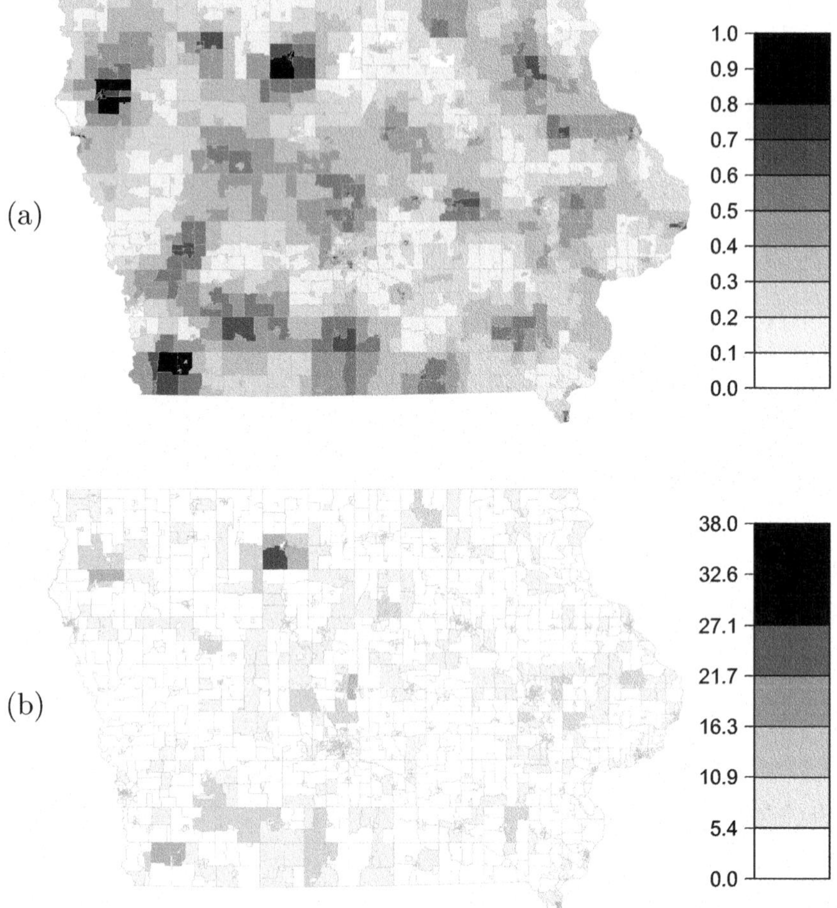

Fig. 2 A single simulated zero-inflated dataset for the census block groups of Iowa. The proportion of renters in each block group was used as a covariate. Panel (**a**) displays the probabilities of incidence. Panel (**b**) displays the ZTP rates, where a given rate is nonzero only if the underlying binary outcome is equal to 1

units occupied by renters (see Sect. 4.4). We used $\boldsymbol{\beta}_1 = (-1, 1)'$ and $\boldsymbol{\beta}_2 = (2, -1)'$. These values for $\boldsymbol{\beta}_1$ indicate that block groups with a high percentage of renters will be more likely to take nonzero values than block groups with a low percentage of renters. The values of $\boldsymbol{\beta}_2$ indicate that as the proportion of renters increases, the ZTP rate decreases. We used 4 and 8 as the diagonal elements of $\boldsymbol{\Sigma}$, and we used five different values for the within-unit correlation: $\rho = 0, 0.2, 0.4, 0.6, 0.8$.

We constructed $\mathbf{Q}_s$ as detailed in Sect. 2.3 and used the same $\mathbf{Q}_s$ for both processes. Eigendecomposition of the Moran operator yielded 1000 basis vectors exhibiting patterns of positive spatial dependence. We used the first $q = 250$ eigenvectors to construct $\mathbf{M}$. This choice of q allowed the responses to exhibit both small- and large-scale spatial variation. We then simulated spatial effects $\boldsymbol{\Delta} = (\boldsymbol{\delta}'_{s1}, \boldsymbol{\delta}'_{s2})'$ from a zero-mean Gaussian distribution with covariance $\boldsymbol{\Sigma} \otimes \mathbf{Q}_s^{-1}$. For the ith block group we simulated y_{i1} from the Bernoulli distribution with success probability $\pi_i = \text{logit}^{-1}(\boldsymbol{x}'_i\boldsymbol{\beta}_1 + \boldsymbol{m}'_i\boldsymbol{\delta}_{s1})$. Conditional on $y_{i1} = 1$, we drew y_{i2} from the zero-truncated Poisson distribution with rate $\lambda_i = \exp(\boldsymbol{x}'_i\boldsymbol{\beta}_2 + \boldsymbol{m}'_i\boldsymbol{\delta}_{s2})$. Finally, we let $y_i = 0$ if $y_{i1} = 0$, or $y_i = y_{i2}$ if $y_{i1} = 1$.

We analyzed 1000 simulated datasets for each of the five correlations. To assess the importance of modeling the dependence within areal units, we applied both the MSAMM and independent SAMMs to each dataset. Key results are shown in Table 1. Extended results, including credible interval coverage rates, are included in an appendix. We see that neglecting within-unit dependence leads to larger biases and, for some parameters, larger mean squared errors, especially for larger values of ρ.

Note that we did not compare the performance of our model to the performance of one or more hurdle MCAR models, for three reasons. First, to the best of our knowledge, no hurdle MCAR model has been implemented in software. Second, any MCAR model is spatially confounded for the same reason that the univariate CAR models are spatially confounded. And third, fitting any MCAR model would be terribly burdensome computationally (with respect to both running time and storage requirement) for the same reason that fitting any univariate CAR model would be burdensome. Hence, pitting a hurdle MCAR model against our hurdle MSAMM would have led to no new knowledge.

4.4 Application of the Hurdle Model to the Iowa Census Data

In this section we apply our areal hurdle model to address deletes from the 2010 US Census within the state of Iowa. Recall that a delete is defined as an address that was deleted from the base count because it did not correspond to a valid housing unit.

Table 1 Results for the simulation study

		MSAMM			Independent SAMMs		
Par.	Truth	Mean est.	Bias	MSE	Mean est.	Bias	MSE
β_{11}	-1	-0.998	0.002	0.006	-0.995	0.005	0.006
β_{12}	1	0.994	-0.006	0.058	0.991	-0.009	0.057
β_{21}	2	2.003	0.003	0.001	2.003	0.003	<0.001
β_{22}	-1	-1.001	-0.001	0.009	-1.001	-0.001	0.009
σ_1^2	4	3.970	-0.030	1.357	3.703	-0.297	1.395
σ_2^2	8	8.274	0.274	0.939	8.153	0.153	0.847
ρ	0	<0.001	<0.001	0.014	–	–	–
β_{11}	-1	-0.998	0.002	0.006	-0.995	0.005	0.006
β_{12}	1	0.994	-0.006	0.058	0.991	-0.009	0.057
β_{21}	2	2.004	0.004	<0.001	2.012	0.012	0.001
β_{22}	-1	-1.000	<0.001	0.008	-1.003	-0.003	0.008
σ_1^2	4	4.009	0.009	1.336	3.703	-0.297	1.395
σ_2^2	8	8.248	0.248	0.933	8.112	0.112	0.838
ρ	0.2	0.201	0.001	0.013	–	–	–
β_{11}	-1	-0.998	0.002	0.006	-0.995	0.005	0.006
β_{12}	1	0.994	-0.006	0.058	0.991	-0.009	0.057
β_{21}	2	2.004	0.004	<0.001	2.020	0.020	0.001
β_{22}	-1	-1.000	<0.001	0.008	-1.004	-0.004	0.008
σ_1^2	4	4.012	0.012	1.345	3.703	-0.297	1.395
σ_2^2	8	8.229	0.229	0.945	8.071	0.071	0.836
ρ	0.4	0.405	0.005	0.012	–	–	–
β_{11}	-1	-0.998	0.002	0.006	-0.995	0.005	0.006
β_{12}	1	0.995	-0.005	0.058	0.991	-0.009	0.057
β_{21}	2	2.005	0.005	<0.001	2.028	0.028	0.002
β_{22}	-1	-1.000	<0.001	0.007	-1.007	-0.007	0.007
σ_1^2	4	4.060	0.060	1.309	3.703	-0.297	1.395
σ_2^2	8	8.238	0.238	0.943	8.018	0.018	0.827
ρ	0.6	0.604	0.004	0.010	–	–	–
β_{11}	-1	-1.000	<0.001	0.006	-0.995	0.005	0.006
β_{12}	1	0.996	-0.004	0.058	0.991	-0.009	0.057
β_{21}	2	2.007	0.007	<0.001	2.036	0.036	0.002
β_{22}	-1	-1.000	<0.001	0.007	-1.008	-0.008	0.007
σ_1^2	4	4.165	0.165	1.263	3.703	-0.297	1.395
σ_2^2	8	8.197	0.197	0.940	7.944	-0.056	0.824
ρ	0.8	0.793	-0.007	0.006	–	–	–

We analyzed 1000 simulated zero-inflated datasets for each value of ρ. MSE denotes mean squared error

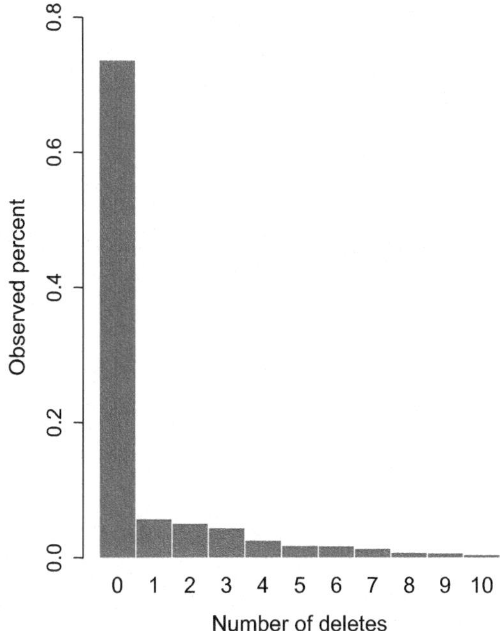

Fig. 3 Histogram of number of address deletes for the 2600 block groups of Iowa from the 2010 Census. Approximately 75% of the outcomes are zeros and 4% of the outcomes are greater than 10

Figure 3 displays a histogram (truncated at 10) showing the zero-inflation present in the data.

In the count component of the model, along with covariates, we introduced an offset calculated via external standardization. Specifically, the offset is the natural log of the total number of addresses in the block group to which a Census form was delivered. We used the same covariates in the binary and count components of the model. Each covariate was a proportion. We used the following covariates and an intercept.

- RURAL_POP: proportion of population living outside of an urban area or urban cluster
- OCCP_HU: proportion of housing units classified as the usual place of residence of the individual or group living there
- RENTER_OCCP_HU: proportion of occupied housing units that are not owner occupied, whether they are rented or occupied without payment of rent
- TEA_MAIL: proportion of addresses that received a Census form in the mail and occupants were instructed to complete and return the form
- FIRST_FRM: proportion of addresses where the first form mailed was completed and returned

As in the simulation study, we used the first 250 eigenvectors of the Moran operator, and we applied both the MSAMM and independent SAMMs.

The results are shown in Table 2. First, we see that the correlation ρ was estimated as 0.73, with a 95% highest posterior density (HPD) interval of $(0.25, 0.91)$.

Table 2 Iowa address delete results for the MSAMM versus independent SAMMs

Predictor/parameter	MSAMM		Independent SAMMs	
	Posterior mean	95% CI	Posterior mean	95% CI
Intercept	7.23	$(-4.43, -9.83)$	7.48	$(-4.93, 10.53)$
RURAL_POP	2.72	$(-2.18, -3.30)$	2.70	$(-2.13, -3.29)$
OCCP_HU	0.68	$(-2.68, -4.60)$	0.25	$(-3.21, -3.61)$
RENTER_OCCP_HU	-3.51	$(-5.73, -1.41)$	-3.60	$(-6.01, -1.16)$
TEA_MAIL	-6.95	$(-7.54, -6.37)$	-6.98	$(-7.60, -6.40)$
FIRST_FRMS	-7.10	$(-9.93, -4.28)$	-6.86	$(-9.82, -3.86)$
σ_1^2	0.67	$(-0.30, -2.04)$	0.94	$(-0.21, -2.52)$
Intercept	3.99	$(-3.50, -4.50)$	3.92	$(-3.07, -4.69)$
RURAL_POP	-0.00	$(-0.18, -0.15)$	0.02	$(-0.14, -0.18)$
OCCP_HU	2.52	$(-1.07, -3.59)$	2.00	$(-0.48, -3.17)$
RENTER_OCCP_HU	-1.47	$(-2.28, -0.68)$	-1.16	$(-1.99, -0.16)$
TEA_MAIL	-2.03	$(-2.22, -1.83)$	-1.98	$(-2.24, -1.73)$
FIRST_FRMS	-6.11	$(-7.39, -4.65)$	-5.50	$(-7.00, -2.93)$
σ_2^2	7.74	$(-4.99, 11.47)$	8.24	$(-5.37, 12.01)$
ρ	0.73	$(-0.25, -0.91)$		–
pD	258.85		172.31	
DIC	-7745		-7735	

Results for the binary components of the models are shown in the top portion of the table. Results for the count components are shown in the bottom portion. CI denotes credible interval

This suggests that the MSAMM, as opposed to independent SAMMs, is appropriate for these data. This claim is further supported by the fact that the correlation model yields a lower deviance information criterion (DIC) value (Spiegelhalter et al. 2002). (A difference in DIC of 10 may not seem substantial but is, in fact, enormous if relative likelihood (Burnham et al. 2011) is used as one's basis for comparison: $\exp(-10/2) = \exp(-5) \approx 0.007$.)

Regarding the regression coefficients for the binary component, we see that all but OCCP_HU have 95% credible intervals that exclude zero. This suggests that, in Iowa, the proportion of occupied housing units is not predictive of the occurrence of deletes within a block group. We see that RENTER_OCCP_HU, TEA_MAIL, and FIRST_FRM offer a "protective effect" against the occurrence of deletes while RURAL_POP is associated with an increased likelihood of deletes. We also see that the variation in the spatial effects for the binary component appears to be small: $\hat{\sigma}_1^2 = 0.67 \, (0.30, 2.04)$.

The regression for the count component tells a somewhat different story. All coefficients save RURAL_POP have 95% credible intervals that exclude zero, suggesting that the proportion of a block group that is rural is not associated with the number of deletes. Similar to the binary component, we see that RENTER_OCCP_HU,

TEA_MAIL, and FIRST_FRM offer a "protective effect" against the number of deletes, but now OCCP_HU is associated with a greater number of deletes. Evidently, the spatial process is not as smooth for the count component since $\hat{\sigma}_2^2 = 7.74\ (4.99, 11.47)$.

The results obtained using our areal hurdle model could be valuable to the Census Bureau. When using the most recent covariate values obtained (such as from official government surveys or administrative records), predictions using our model could help to characterize deletes, which is an indication of demographic change or stability in an area. The spatial component of our model can assist in designing efficient and cost-effective address updating operations. For example, it can inform Census Bureau personnel as to *clusters* of block groups that are candidates for updating in a non-decennial Census setting. Focusing on adjacent regions within a cluster will be advantageous over assessing sets of block groups that might have only "stable" block groups as neighbors. Such clustering will not always be captured accurately by non-spatial zero-inflated models.

5   Discussion and Future Work

Our proposed methods for handling multivariate and zero-inflated areal data offer improved regression inference while greatly reducing computing time and storage requirements. Our simulation study illustrates the benefit of accounting for dependence within areal units as well as among areal units. This is not surprising: in general, multivariate data call for multivariate methods.

The count distribution used for our model is the Poisson, which requires the assumption of equi-dispersion. Of course, the data could be heavily over- or under-dispersed, in which case other distributions could be developed in our MSAMM setup, such as the negative binomial or the Conway–Maxwell–Poisson distribution. Both of these distributions have an additional parameter that characterizes the dispersion, which could possibly depend on spatially varying covariates. These different models would be novel, but would require additional numerical work to demonstrate how well they improve the fits.

Application of our areal Poisson hurdle model to zero-inflated Census data provided a superior fit relative to that provided by independent univariate models, at no extra computational cost. Most importantly, our methodology provides a compelling framework for understanding dynamic features of the USA, which could aid the planning of various Census Bureau operations. Moreover, our methodology could be extended to handle additional data challenges faced by the Census Bureau. For example, a spatiotemporal extension of our MSAMM could be useful for analyzing data from historical databases being developed by the Census Bureau. In such a model, time-dependent covariates could be viewed as driving the deletion of housing units.

Appendix: Supplementary Materials

Multivariate Spatial Effect Reparameterization

For the multivariate sparse areal mixed model (MSAMM), when the design matrices are the same across multivariate outcomes, i.e., $\mathbf{X}_1 = \mathbf{X}_2 = \cdots = \mathbf{X}_J$, the first and second stages can be written as

$$g_j \left\{ \mathbb{E} \left(\mathbf{y}_j \mid \boldsymbol{\beta}_j, \, \boldsymbol{\delta}_{sj} \right) \right\} = \mathbf{X}\boldsymbol{\beta}_j + \mathbf{M}\boldsymbol{\delta}_{sj} \qquad (j = 1, \ldots, J)$$

$$p \left(\boldsymbol{\Delta} \mid \boldsymbol{\Sigma} \right) = \mathcal{N} \left(\mathbf{0}, \, \boldsymbol{\Sigma} \otimes \mathbf{Q}_s^{-1} \right),$$

where $\boldsymbol{\Delta} = \left(\boldsymbol{\delta}_{s1}', \ldots, \boldsymbol{\delta}_{sJ}' \right)'$, each $\boldsymbol{\delta}_{sj}$ is $q \times 1$, $\boldsymbol{\Sigma}$ is the $J \times J$ covariance matrix, and $\mathbf{Q}_s$ is the $q \times q$ spatial precision matrix.

Computation can be eased considerably as follows. Let $\mathbf{R}_s$ be the upper Cholesky triangle of $\mathbf{Q}_s$, and let $\mathbf{W}_s = \mathbf{R}_s^{-1}$ such that $\mathbf{W}_s \mathbf{W}_s' = \mathbf{Q}_s^{-1}$. Then, for $\boldsymbol{\Psi} = \left(\boldsymbol{\psi}_{s1}', \ldots, \boldsymbol{\psi}_{sJ}' \right)'$, each $\boldsymbol{\psi}_{sj}$ is $q \times 1$, and $\boldsymbol{\Psi} \mid \boldsymbol{\Sigma} \sim \mathcal{N} \left(\mathbf{0}, \, \boldsymbol{\Sigma} \otimes \mathbf{I}_q \right)$, we have that $\left(\mathbf{I}_J \otimes \mathbf{W}_s \right) \boldsymbol{\Psi}$ and $\boldsymbol{\Delta}$ have the same distribution conditional on $\boldsymbol{\Sigma}$. This is easy to see since $\mathbb{E} \left\{ \left(\mathbf{I}_J \otimes \mathbf{W}_s \right) \boldsymbol{\Psi} \right\} = \left(\mathbf{I}_J \otimes \mathbf{W}_s \right) \mathbb{E} \left(\boldsymbol{\Psi} \right) = \mathbf{0}$ and

$$\mathrm{cov} \left\{ \left(\mathbf{I}_J \otimes \mathbf{W}_s \right) \boldsymbol{\Psi} \right\} = \left(\mathbf{I}_J \otimes \mathbf{W}_s \right) \left(\boldsymbol{\Sigma} \otimes \mathbf{I}_q \right) \left(\mathbf{I}_J \otimes \mathbf{W}_s \right)'$$

$$= \boldsymbol{\Sigma} \otimes \mathbf{Q}_s^{-1}.$$

Hence, the model's first and second stages can now be written as

$$g_j \left\{ \mathbb{E} \left(\mathbf{y}_j \mid \boldsymbol{\beta}_j, \, \boldsymbol{\psi}_{sj} \right) \right\} = \mathbf{X}\boldsymbol{\beta}_j + \mathbf{M}\mathbf{W}_s \boldsymbol{\psi}_{sj} \qquad (j = 1, \ldots, J)$$

$$p \left(\boldsymbol{\Psi} \mid \boldsymbol{\Sigma} \right) = \mathcal{N} \left(\mathbf{0}, \, \boldsymbol{\Sigma} \otimes \mathbf{I}_q \right).$$

Now suppose that $\mathbf{X}_1 \neq \mathbf{X}_2 \neq \cdots \neq \mathbf{X}_J$. Then we have

$$g_j \left\{ \mathbb{E} \left(\mathbf{y}_j \mid \boldsymbol{\beta}_j, \, \boldsymbol{\delta}_{sj} \right) \right\} = \mathbf{X}_j \boldsymbol{\beta}_j + \mathbf{M}_j \boldsymbol{\delta}_{sj}$$

$$p \left(\boldsymbol{\Delta} \mid \boldsymbol{\Sigma} \right) = \mathcal{N} \left[\mathbf{0}, \, \left\{ \mathbf{R}' \left(\boldsymbol{\Sigma}^{-1} \otimes \mathbf{I}_q \right) \mathbf{R} \right\}^{-1} \right],$$

where $\boldsymbol{\Delta} = \left(\boldsymbol{\delta}_{s1}', \ldots, \boldsymbol{\delta}_{sJ}' \right)'$, $\mathbf{R} = \mathrm{bdiag} \left(\mathbf{R}_{s1}, \ldots, \mathbf{R}_{sJ} \right)$, and $\mathbf{R}_{sj}' \mathbf{R}_{sj} = \mathbf{Q}_{sj}$, where $\mathbf{R}_{sj}$ is the upper Cholesky triangle of $\mathbf{Q}_{sj}$. For ease of exposition, let $J = 2$ (the following easily extends to the case when $J > 2$). The prior distribution of the spatial effects can be written

$$\begin{pmatrix} \boldsymbol{\delta}_{s1} \\ \boldsymbol{\delta}_{s2} \end{pmatrix} \mid \boldsymbol{\Sigma} \ \sim \ \mathcal{N}\left[\begin{pmatrix} \mathbf{0} \\ \mathbf{0} \end{pmatrix}, \ \left\{ \begin{pmatrix} \mathbf{R}_{s1} & \mathbf{0} \\ \mathbf{0} & \mathbf{R}_{s2} \end{pmatrix}' \left(\boldsymbol{\Sigma}^{-1} \otimes \mathbf{I}_q \right) \begin{pmatrix} \mathbf{R}_{s1} & \mathbf{0} \\ \mathbf{0} & \mathbf{R}_{s2} \end{pmatrix} \right\}^{-1} \right]$$

$$= \ \mathcal{N}\left[\begin{pmatrix} \mathbf{0} \\ \mathbf{0} \end{pmatrix}, \ \begin{pmatrix} \mathbf{W}_{s1} & \mathbf{0} \\ \mathbf{0} & \mathbf{W}_{s2} \end{pmatrix} \left(\boldsymbol{\Sigma} \otimes \mathbf{I}_q \right) \begin{pmatrix} \mathbf{W}_{s1} & \mathbf{0} \\ \mathbf{0} & \mathbf{W}_{s2} \end{pmatrix}' \right],$$

where $\mathbf{W}_{sj} = \mathbf{R}_{sj}^{-1}$ $(j = 1, 2)$, and we have used the fact that $\left(\mathbf{R}_{sj}^{-1} \right)' = \left(\mathbf{R}_{sj}' \right)^{-1}$. Now, suppose we have

$$\begin{pmatrix} \boldsymbol{\psi}_{s1} \\ \boldsymbol{\psi}_{s2} \end{pmatrix} \mid \boldsymbol{\Sigma} \ \sim \ \mathcal{N}\left\{ \begin{pmatrix} \mathbf{0} \\ \mathbf{0} \end{pmatrix}, \ \boldsymbol{\Sigma} \otimes \mathbf{I}_q \right\}.$$

Using basic properties of the multivariate normal distribution, we have that

$$\begin{pmatrix} \mathbf{W}_{s1} & \mathbf{0} \\ \mathbf{0} & \mathbf{W}_{s2} \end{pmatrix} \begin{pmatrix} \boldsymbol{\psi}_{s1} \\ \boldsymbol{\psi}_{s2} \end{pmatrix} \mid \boldsymbol{\Sigma} \ \sim \ \mathcal{N}\left\{ \begin{pmatrix} \mathbf{0} \\ \mathbf{0} \end{pmatrix}, \ \begin{pmatrix} \mathbf{W}_{s1} & \mathbf{0} \\ \mathbf{0} & \mathbf{W}_{s2} \end{pmatrix} \left(\boldsymbol{\Sigma} \otimes \mathbf{I}_q \right) \begin{pmatrix} \mathbf{W}_{s1} & \mathbf{0} \\ \mathbf{0} & \mathbf{W}_{s2} \end{pmatrix}' \right\}.$$

Then, since

$$\begin{pmatrix} \mathbf{W}_{s1} & \mathbf{0} \\ \mathbf{0} & \mathbf{W}_{s2} \end{pmatrix} \begin{pmatrix} \boldsymbol{\psi}_{s1} \\ \boldsymbol{\psi}_{s2} \end{pmatrix} = \begin{pmatrix} \mathbf{W}_{s1}\boldsymbol{\psi}_{s1} \\ \mathbf{W}_{s2}\boldsymbol{\psi}_{s2} \end{pmatrix},$$

we can apply a reparameterization similar to the case where design matrices are equivalent across the outcomes. Thus we can specify the first and second stages of the model as

$$g_j\left\{ \mathbb{E}\left(\boldsymbol{y}_j \mid \boldsymbol{\beta}_j, \ \boldsymbol{\psi}_{sj} \right) \right\} = \mathbf{X}_j \boldsymbol{\beta}_j + \mathbf{M}_j \mathbf{W}_{sj} \boldsymbol{\psi}_{sj}$$

$$p\left(\boldsymbol{\Psi} \mid \boldsymbol{\Sigma} \right) = \mathcal{N}\left(\mathbf{0}, \ \boldsymbol{\Sigma} \otimes \mathbf{I}_q \right).$$

Extended Simulation Results

Table 3 provides complete results for our simulation study.

Table 3 Extended results for our simulation study

Par.	Truth	MSAMM				Independent SAMMs			
		Mean est.	Bias (%)	MSE	Coverage	Mean est.	Bias (%)	MSE	Coverage
β_{11}	-1	-0.998	0.002 (0.002)	0.006	0.948	-0.995	0.005 (0.005)	0.006	0.942
β_{12}	1	0.994	-0.006 (0.006)	0.058	0.930	0.991	-0.009 (0.009)	0.057	0.931
β_{21}	2	2.003	0.003 (0.001)	0.001	0.953	2.003	0.003 (0.002)	<0.001	0.962
β_{22}	-1	-1.001	-0.001 (0.001)	0.009	0.958	-1.001	-0.001 (0.001)	0.009	0.955
σ_1^2	4	3.970	-0.030 (0.008)	1.357	0.945	3.703	-0.297 (0.074)	1.395	0.906
σ_2^2	8	8.274	0.274 (0.034)	0.939	0.975	8.153	0.153 (0.019)	0.847	0.977
ρ	0	<0.001	<0.001 ($-$)	0.014	0.952	$-$	$-$	$-$	$-$
β_{11}	-1	-0.998	0.002 (0.002)	0.006	0.945	-0.995	0.005 (0.005)	0.006	0.942
β_{12}	1	0.994	-0.006 (0.006)	0.058	0.929	0.991	-0.009 (0.009)	0.057	0.931
β_{21}	2	2.004	0.004 (0.002)	<0.001	0.959	2.012	0.012 (0.006)	0.001	0.944
β_{22}	-1	-1.000	<0.001 (<0.001)	0.008	0.953	-1.003	-0.003 (0.003)	0.008	0.949
σ_1^2	4	4.009	0.009 (0.002)	1.336	0.950	3.703	-0.297 (0.074)	1.395	0.906
σ_2^2	8	8.248	0.248 (0.031)	0.933	0.970	8.112	0.112 (0.014)	0.838	0.974
ρ	0.2	0.201	0.001 (0.006)	0.013	0.954	$-$	$-$	$-$	$-$
β_{11}	-1	-0.998	0.002 (0.002)	0.006	0.943	-0.995	0.005 (0.005)	0.006	0.942
β_{12}	1	0.994	-0.006 (0.006)	0.058	0.929	0.991	-0.009 (0.009)	0.057	0.931
β_{21}	2	2.004	0.004 (0.002)	<0.001	0.962	2.020	0.020 (0.010)	0.001	0.930
β_{22}	-1	-1.000	<0.001 (<0.001)	0.008	0.954	-1.004	-0.004 (0.004)	0.008	0.951
σ_1^2	4	4.012	0.012 (0.003)	1.345	0.949	3.703	-0.297 (0.074)	1.395	0.906
σ_2^2	8	8.229	0.229 (0.029)	0.945	0.970	8.071	0.071 (0.009)	0.836	0.966
ρ	0.4	0.405	0.005 (0.011)	0.012	0.949	$-$	$-$	$-$	$-$
β_{11}	-1	-0.998	0.002 (0.002)	0.006	0.945	-0.995	0.005 (0.005)	0.006	0.942

β_{12}	1	0.995	−0.005 (0.005)	0.058	0.932	0.991	−0.009 (0.009)	0.057	0.931
β_{21}	2	2.005	0.005 (0.003)	<0.001	0.966	2.028	0.028 (0.014)	0.002	0.903
β_{22}	−1	−1.000	<0.001 (<0.001)	0.007	0.957	−1.007	−0.007 (0.007)	0.007	0.950
σ_1^2	4	4.060	0.060 (0.015)	1.309	0.940	3.703	−0.297 (0.074)	1.395	0.906
σ_2^2	8	8.238	0.238 (0.030)	0.943	0.976	8.018	0.018 (0.002)	0.827	0.973
ρ	0.6	0.604	0.004 (0.006)	0.010	0.945	−	−	−	−
β_{11}	−1	−1.000	<0.001 (<0.001)	0.006	0.945	−0.995	0.005 (0.005)	0.006	0.942
β_{12}	1	0.996	−0.004 (0.004)	0.058	0.933	0.991	−0.009 (0.009)	0.057	0.931
β_{21}	2	2.007	0.007 (0.003)	<0.001	0.963	2.036	0.036 (0.018)	0.002	0.853
β_{22}	−1	−1.000	<0.001 (<0.001)	0.007	0.959	−1.008	−0.008 (0.008)	0.007	0.946
σ_1^2	4	4.165	0.165 (0.041)	1.263	0.945	3.703	−0.297 (0.074)	1.395	0.906
σ_2^2	8	8.197	0.197 (0.025)	0.940	0.979	7.944	−0.056 (0.007)	0.824	0.965
ρ	0.8	0.793	−0.007 (0.009)	0.006	0.953	−	−	−	−

References

Agarwal, D.K., Gelfand, A.E., Citron-Pousty, S.: Zero-inflated models with application to spatial count data. Environ. Ecol. Stat. **9**(4), 341–355 (2002)

Alfó, M., Nieddu, L., Vicari, D.: Finite mixture models for mapping spatially dependent disease counts. Biom. J. **51**(1), 84–97 (2009). http://dx.doi.org/10.1002/bimj.200810494

Assunção, R., Krainski, E.: Neighborhood dependence in Bayesian spatial models. Biom. J. **51**(5), 851–869 (2009)

Barnard, J., McCulloch, R., Meng, X.L.: Modeling covariance matrices in terms of standard deviations and correlations, with application to shrinkage. Stat. Sin. **10**(4), 1281–1312 (2000)

Besag, J., Kooperberg, C.: On conditional and intrinsic autoregression. Biometrika **82**(4), 733–746 (1995)

Boots, B., Tiefelsdorf, M.: Global and local spatial autocorrelation in bounded regular tessellations. J. Geogr. Syst. **2**(4), 319 (2000)

Boucher, J.P., Denuit, M., Guillen, M.: Number of accidents or number of claims? An approach with zero-inflated Poisson models for panel data. J. Risk Insur. **76**(4), 821–846 (2009)

Bradley, J.R., Holan, S.H., Wikle, C.K.: Multivariate spatio-temporal models for high-dimensional areal data with application to longitudinal employer-household dynamics. Ann. Appl. Stat. **9**(4), 1761–1791 (2015)

Burnham, K.P., Anderson, D.R., Huyvaert, K.P.: AIC model selection and multimodel inference in behavioral ecology: some background, observations, and comparisons. Behav. Ecol. Sociobiol. **65**(1), 23–35 (2011)

Carlin, B.P., Banerjee, S.: Hierarchical multivariate CAR models for spatio-temporally correlated survival data (with discussion). In: Bayarri, M., Berger, J., Bernardo, J., Dawid, A., Heckerman, D., Smith, A., West, M. (eds.), Bayesian Statistics 7, pp. 45–63. Oxford University Press, New York (2003)

Clayton, D., Bernardinelli, L., Montomoli, C.: Spatial correlation in ecological analysis. Int. J. Epidemiol. **22**(6), 1193–1202 (1993)

Cohen, A.C.: Estimating the parameter in a conditional Poisson distribution. Biometrics **16**(2), 203–211 (1960)

Cook, T., Norwood, J., Cork, D., Panel to Review the 2010 Census, Committee on National Statistics, Division of Behavioral and Social Sciences and Education, National Research Council: Change and the 2020 Census: Not Whether But How. National Academies Press, Washington, D.C. (2011)

Donoho, D.L., Elad, M.: Optimally sparse representation in general (nonorthogonal) dictionaries via ℓ^1 minimization. Proc. Natl. Acad. Sci. **100**(5), 2197–2202 (2003)

Eddelbuettel, D., Francois, R.: Rcpp: Seamless R and C++ integration. J. Stat. Softw. **40**(8), 1–18 (2011)

Eddelbuettel, D., Sanderson, C.: RcppArmadillo: Accelerating R with high-performance C++ linear algebra. Comput. Stat. Data Anal. **71**, 1054–1063 (2014)

Flegal, J.M., Haran, M., Jones, G.L.: Markov chain Monte Carlo: can we trust the third significant figure? Stat. Sci. 23(2), 250–260 (2008)

Gelfand, A.E., Vounatsou, P.: Proper multivariate conditional autoregressive models for spatial data analysis. Biostatistics **4**(1), 11–15 (2003)

Green, P.J., Richardson, S.: Hidden Markov models and disease mapping. J. Am. Stat. Assoc. **97**(460), 1055–1070 (2002). https://doi.org/10.1198/016214502388618870

Griffith, D.A.: Spatial Autocorrelation and Spatial Filtering: Gaining Understanding Through Theory and Scientific Visualization. Springer, Berlin (2003)

Haran, M., Hughes, J.: batchmeans: consistent batch means estimation of Monte Carlo standard errors. Denver (2016)

Haran, M., Hodges, J., Carlin, B.: Accelerating computation in Markov random field models for spatial data via structured MCMC. J. Comput. Graph. Stat. **12**(2), 249–264 (2003)

Haran, M., Tierney, L.: On automating Markov chain Monte Carlo for a class of spatial models. Preprint (2012). arXiv:12050499

Hodges, J., Reich, B.: Adding spatially-correlated errors can mess up the fixed effect you love. Am. Stat. **64**(4), 325–334 (2010)

Huang, A., Wand, M.: Simple marginally noninformative prior distributions for covariance matrices. Bayesian Anal. **8**(2), 439–452 (2013)

Hughes, J., Haran, M.: Dimension reduction and alleviation of confounding for spatial generalized linear mixed models. J. R. Stat. Soc. Ser. B Stat. Methodol. 75(1), 139–159 (2013)

Ihaka, R., Gentleman, R.: R: a language for data analysis and graphics. J. Comput. Graph. Stat. **5**, 299–314 (1996)

Jin, X., Carlin, B.P., Banerjee, S.: Generalized hierarchical multivariate CAR models for areal data. Biometrics **61**(4), 950–961 (2005)

Knorr-Held, L., Rue, H.: On block updating in Markov random field models for disease mapping. Scand. J. Stat. **29**(4), 597–614 (2002)

Lambert, D.: Zero-inflated Poisson regression, with an application to defects in manufacturing. Technometrics **34**(1), 1–14 (1992)

Leroux, B.G., Lei, X., Breslow, N.: Estimation of disease rates in small areas: a new mixed model for spatial dependence. Inst. Math. Appl. **116**, 179–191 (2000)

Lewandowski, D., Kurowicka, D., Joe, H.: Generating random correlation matrices based on vines and extended onion method. J. Multivar. Anal. **100**(9), 1989–2001 (2009)

Martinez-Beneito, M.A.: A general modelling framework for multivariate disease mapping. Biometrika **100**(3), 539–553 (2013)

Moran, P.: Notes on continuous stochastic phenomena. Biometrika **37**(1/2), 17–23 (1950)

Neelon, B., Ghosh, P., Loebs, P.F.: A spatial Poisson hurdle model for exploring geographic variation in emergency department visits. J. R. Stat. Soc. Ser. A Stat. Soc. **176**(2), 389–413 (2013)

Neelon, B., Zhu, L., Neelon, S.E.B.: Bayesian two-part spatial models for semicontinuous data with application to emergency department expenditures. Biostatistics **16**(3), 465–479 (2015)

Qiu, Y.: Spectra: sparse eigenvalue computation toolkit as a redesigned ARPACK. http://spectralib. org (2017)

Rathbun, S.L., Fei, S.: A spatial zero-inflated Poisson regression model for oak regeneration. Environ. Ecol. Stat. **13**(4):409–426 (2006)

Recta, V., Haran, M., Rosenberger, J.L.: A two-stage model for incidence and prevalence in point-level spatial count data. Environmetrics **23**(2), 162–174 (2012)

Reich, B., Hodges, J., Zadnik, V.: Effects of residual smoothing on the posterior of the fixed effects in disease-mapping models. Biometrics **62**(4), 1197–1206 (2006)

Sanderson, C.: Armadillo: an open source C++ linear algebra library for fast prototyping and computationally intensive experiments. Technical Report; NICTA (2010)

Singh, J.: A characterization of positive Poisson distribution and its statistical application. SIAM J. Appl. Math. **34**(3), 545–548 (1978)

Spiegelhalter, D.J., Best, N.G., Carlin, B.P., Van Der Linde, A.: Bayesian measures of model complexity and fit. J. R. Stat. Soc. Ser. B Stat. Methodol. **64**(4), 583–639 (2002)

Stroustrup, B.: The C++ Programming Language. Pearson Education, New Jersey (2013)

Tiefelsdorf, M., Griffith, D.A.: Semiparametric filtering of spatial autocorrelation: the eigenvector approach. Environ. Plan. A **39**(5), 1193 (2007)

Torabi, M.: Hierarchical multivariate mixture generalized linear models for the analysis of spatial data: an application to disease mapping. Biom. J. **58**(5), 1138–1150 (2016)

U.S. Census Bureau: 2020 Census operational plan: a new design for the 21st century (2015)

Ver Hoef, J.M., Jansen, J.K.: Space-time zero-inflated count models of harbor seals. Environmetrics **18**(7), 697–712 (2007)

Wall, M.: A close look at the spatial structure implied by the CAR and SAR models. J. Stat. Plan. Inference **121**(2), 311–324 (2004)

Wikle, C.K., Anderson, C.J.: Climatological analysis of tornado report counts using a hierarchical Bayesian spatiotemporal model. J. Geophys. Res. Atmos. (1984–2012) **108**(D24), 1–15 (2003). https://doi.org/10.1029/2002JD002806

Young, D.S., Raim, A.M., Johnson, N.R.: Zero-inflated modelling for characterizing coverage errors of extracts from the US Census Bureau's Master Address File. J. R. Stat. Soc. Ser. A Stat. Soc. **180**(1), 73–97 (2017)

Wavelet Kernels for Support Matrix Machines

Edgard M. Maboudou-Tchao

1 Introduction

Binary classification is a supervised machine learning technique that classifies the elements of a certain set into two groups based on a classification rule. Classification problems occur in different areas such as text categorization, web page classification, news classification, spam mail filtering, and discriminating tumorous genes from non-tumorous genes. Support vector machine (SVM) (Cortes and Vapnik 1995) is a powerful method based on statistical learning theory and has been proven to perform better than existing methods in many aspects.

Typical classification methods use vectors as input samples. In the case that input samples are second-order tensors or matrices, they need to be vectorized first before applying classical classification techniques. This practice will destroy the structure information of the data matrix as well as the correlation between the variables. Also, the dimension of the resulting matrix vectorized will be large and this can cause a dimensionality issue. Consequently, computation time will increase significantly. Support matrix machine was introduced for binary classification of matrices (Shi and Zhang 2009). Luo et al. (2015) proposed a penalized support matrix machine by using a spectral elastic net regularization, which combines the Frobenius norm and nuclear norm to constrain the regression matrix. Maboudou-Tchao (2017) proposed support matrix data description (SMDD) for one-class classification of matrices. Xia and Fan (2016) suggested a penalized least squares support matrix based on bilevel programming by using both the Frobenius norm and nuclear norm. Zheng et al. (2017) proposed sparse support matrix by regularizing a combination of nuclear norm and ℓ_1 norm of the regression matrix.

E. M. Maboudou-Tchao (⊠)
Department of Statistics, University of Central Florida, Orlando, FL, USA
e-mail: edgard.maboudou@ucf.edu

© Springer Nature Switzerland AG 2019
N. Diawara (ed.), *Modern Statistical Methods for Spatial and Multivariate Data,*
STEAM-H: Science, Technology, Engineering, Agriculture, Mathematics & Health,
https://doi.org/10.1007/978-3-030-11431-2_4

In case of nonlinear classification problem, the input space is mapped into a higher dimensional feature space so that the problem becomes linearly separable. However, the mapping function does not have to be known. Instead, SVMs use the kernel trick, which actually inverts the chain of processes by choosing a kernel rather than choosing a mapping first. A necessary and sufficient condition for the kernel to be a valid kernel is to satisfy the condition given by Mercer's theorem (Mercer 1909). Many Mercer kernels are available, including Gaussian and polynomial kernels. Recent studies proposed Mercer kernels based on wavelet techniques. Zhang et al. (2004) used Morlet wavelet kernel as kernel for SVM. Wu and Zhao (2006) also used Morlet wavelet kernel for least squares SVM.

Support matrix machine (SMM) is a matrix version of SVM and is based on the matrix space. It accepts directly a matrix as inputs without the need of vectorization. By constructing a classifier in the matrix space, the data structure information is retained and it helps overcome the overfitting problem encountered mostly in vector-based learning. This paper will discuss first support matrix machine without using regularization, support matrix regression, and then extend wavelet kernels of SVMs to wavelet kernels for support matrix machines. A new support matrix machine learning algorithm can therefore be built using these new wavelet kernels for matrices.

2　Overview of Support Matrix Machine (SMM)

The standard support vector machines (SVM) aim at finding the optimal hyper-plane that maximizes the margin between two classes. SVM are solved using quadratic programming methods. SVM are easily extended to accept matrix as input. For the two-class classification problem, let the training set be $D = ((\mathbf{X}_1, y_1), \dots, (\mathbf{X}_N, y_N)) \in (\mathcal{X} \times \{-1, 1\})^N$, where N is the number of matrices and $\mathcal{X} \subseteq \mathbb{R}^n \otimes \mathbb{R}^p$ is an original matrix input space. $\mathbb{R}^n$ and $\mathbb{R}^p$ are two vector spaces. SMM consists in solving the following primal problem:

$$
\begin{aligned}
\underset{\mathbf{W}}{\text{minimize}} \quad & \frac{1}{2} tr(\mathbf{W}'\mathbf{W}) + C \sum_{j=1}^{N} \xi_j, \\
\text{subject to} \quad & y_j \left(tr(\mathbf{W}'\varphi(\mathbf{X}_j)) + b \right) \geq 1 - \xi_j, \quad j = 1, 2, \dots, N \\
& \xi_j \geq 0, \quad j = 1, 2, \dots, N.
\end{aligned}
\tag{1}
$$

where $\mathbf{W} \in R^{n \times p}$ is the matrix of regression coefficients, $tr(.)$ is the trace operator, ξ_j are the slack variables, the parameter $C > 0$ is introduced to control the influence of the slack variables, and φ is a function mapping data to a higher dimensional Hilbert space.

This optimization problem is convex with respect to the primal variables $\mathbf{W}$, b, and $\boldsymbol{\xi}$ since each of the $f_0, f_1, \ldots, f_N$ are convex where

$$f_0(\mathbf{W}, b, \boldsymbol{\xi}) = \frac{1}{2} tr(\mathbf{W}'\mathbf{W}) + C \sum_{j=1}^{N} \xi_j$$

and

$$f_j(\mathbf{W}, b, \boldsymbol{\xi}) = 1 - \xi_j - y_j \left(tr(\mathbf{W}'\varphi(\mathbf{X}_j)) + b \right) \leq 0, \ j = 1, \ldots, N.$$

Next, since the inequality constraints are affine and there always exist some $\xi_j \geq 0$ and $y_j \left(tr(\mathbf{W}'\varphi(\mathbf{X}_j)) + b \right) \geq 1 - \xi_j$, Slater's conditions are met and thus strong duality holds. It follows that the duality gap is zero and the optimal values of the primal and dual problems are equal. Consequently, it ensures that the original problem (primal) can be solved through the Lagrange dual problem, which is usually easier to solve than the primal.

The Lagrangian of the given problem is

$$
\begin{aligned}
L(\mathbf{W}, b, \boldsymbol{\alpha}, \boldsymbol{\xi}) &= \frac{1}{2} tr(\mathbf{W}'\mathbf{W}) + C \sum_{j=1}^{N} \xi_j - \sum_{j=1}^{N} \alpha_j \left[y_j \left(tr(\mathbf{W}'\varphi(\mathbf{X}_j)) + b \right) - 1 + \xi_j \right] - \sum_{j=1}^{N} \gamma_j \xi_j \\
&= \frac{1}{2} tr(\mathbf{W}'\mathbf{W}) + \sum_{j=1}^{N} \alpha_j \left(1 - y_j \left(tr(\mathbf{W}'\varphi(\mathbf{X}_j)) + b \right) \right) + \sum_{j=1}^{N} \xi_j \left(C - \alpha_j - \gamma_j \right),
\end{aligned}
$$
$$(2)$$

where α_j and γ_j are positive Lagrange multipliers. To construct the dual problem, we need to determine the optimal $\mathbf{W}$, $\boldsymbol{\xi}$, and b in terms of the dual variables. We achieve this by differentiating the Lagrangian with respect to the primal variables.

$$\frac{\partial L}{\partial \mathbf{W}} = 0 \implies \mathbf{W} - \sum_{j=1}^{N} \alpha_j y_j \varphi(\mathbf{X}_j) = 0,$$

$$\frac{\partial L}{\partial b} = 0 \implies \sum_{j=1}^{N} \alpha_j y_j = 0,$$

$$\frac{\partial L}{\partial \xi_j} = 0 \implies C - \alpha_j - \beta_j = 0,$$

So, the key takeaways are

1. $\mathbf{W} = \sum_{j=1}^{N} \alpha_j y_j \varphi(\mathbf{X}_j)$,
2. $\sum_{j=1}^{N} \alpha_j y_j = 0$,
3. $\beta_j = C - \alpha_j$

It follows now that the Kuhn–Karush–Tucker (KKT) conditions are

(i) Primal feasibility:

$$1 - \xi_j - y_j \left(tr(\mathbf{W}'\varphi(\mathbf{X}_j)) + b \right) \geq 0$$

and

$$\xi_j \geq 0,$$

(ii) Dual feasibility:

$$\alpha_j \geq 0$$

and

$$\gamma_j \geq 0,$$

(iii) Complementary slackness:

$$\alpha_j \left[1 - \xi_j - y_j \left(tr(\mathbf{W}'\varphi(\mathbf{X}_j)) + b \right) \right] = 0$$

and

$$\gamma_j \xi_j = 0.$$

Now, from the first equation of the complementary slackness condition, the objects for which $\alpha_j = 0$ are not on the margin and do not impact the value of $\mathbf{W}$. On the other hand, the objects for which $\alpha_j > 0$ do impact the value of $\mathbf{W}$. These matrices $\mathbf{X}_j$ corresponding to $\alpha_j > 0$ are the support matrices. The support matrices that correspond to matrices located on the decision boundary, with $0 < \alpha_j < C$, are the margin support matrices. The other support matrices, with $\alpha_j = C$, are the non-margin support matrices.

The next step is to maximize the dual problem. Plugging $\mathbf{W}$ into the Lagrangian L and taking into account that $\beta_j = C - \alpha_j$, the dual problem becomes

$$\underset{\alpha}{\text{Maximize}} \quad \sum_{j=1}^{N} \alpha_j - \frac{1}{2} \sum_{i,j=1}^{N} y_i y_j \alpha_i \alpha_j tr(\varphi(\mathbf{X}_i)'\varphi(\mathbf{X}_j)),$$

$$\text{subject to} \quad 0 \leq \alpha_j \leq C \quad \forall j, \tag{3}$$

$$\sum_{j=1}^{N} \alpha_j y_j = 0.$$

Now, if we let $K(\mathbf{X}, \mathbf{Y})$ represent the inner product $tr(\varphi(\mathbf{X})'\varphi(\mathbf{Y}))$ in a higher dimensional space, the dual problem becomes

$$\underset{\alpha}{\text{Maximize}} \quad \sum_{j=1}^{N} \alpha_j - \frac{1}{2} \sum_{i,j=1}^{N} y_i y_j \alpha_i \alpha_j K(\mathbf{X}_i, \mathbf{X}_j),$$

$$\text{subject to} \quad 0 \le \alpha_j \le C \quad \forall j, \tag{4}$$

$$\sum_{j=1}^{N} \alpha_j y_j = 0.$$

The optimization problem is formulated as a strongly convex quadratic problem (QP) whose dual is also a QP. This can be solved easily using any quadratic programming software.

Once the optimal values α_j^* are obtained, the optimal matrix $\mathbf{W}^*$ is found by plugging α_j^* into the equation of $\mathbf{W}$, that is

$$\mathbf{W}^* = \sum_{j=1}^{N} \alpha_j^* y_j \varphi(\mathbf{X}_j). \tag{5}$$

The next step is to evaluate the offset b. b can be found by using a support matrix $\mathbf{X}_j$ and the complementary slackness condition. Alternatively, this can also be achieved using the set of all support matrices by finding an average over all support matrices as

$$b = \frac{1}{N_s} \sum_{i \in S} \left(y_i - \sum_{j \in S} \alpha_j^* y_j K(\mathbf{X}_i, \mathbf{X}_j) \right). \tag{6}$$

Each new matrix point $\mathbf{X}_0$ is classified by evaluating

$$y_0 = sgn\left(\sum_{i-1}^{N} \alpha_i y_i tr\left(\varphi(\mathbf{X}_i)'\varphi(\mathbf{X}_0)\right) + b \right) = sgn\left(\sum_{i-1}^{N} \alpha_i y_i K(\mathbf{X}_i, \mathbf{X}_0) + b \right). \tag{7}$$

3 Support Matrix Regression (SMR)

Support matrix machines can easily be applied to matrix regression problems by using an alternative loss function (Smola 1996). The key for that loss function is to have a distance measure. The commonly used loss functions are the quadratic loss function based on the conventional least squares error criterion, Laplacian loss function that is less sensitive to outliers than the quadratic loss function, polynomial loss function, piecewise polynomial loss function, Huber loss function that is a robust loss function with optimal properties when the underlying distribution of the data is unknown, and the ϵ-insensitive loss function. The first three loss functions do not produce sparseness in the support vectors. To address that issue (Vapnik 1995) suggested the ϵ-insensitive loss function as an approximation to Huber's loss function that enables a sparse set of support vectors to be obtained. The ϵ-insensitive loss function is defined as:

$$L_\epsilon(y) = \begin{cases} 0 & \text{if} \quad |f(x) - y| < \epsilon \\ |f(x) - y| - \epsilon & \text{otherwise.} \end{cases} \tag{8}$$

3.1 Basic Idea

Let the training set be $D = ((\mathbf{X}_1, y_1), \ldots, (\mathbf{X}_N, y_N)) \in (\mathcal{X} \times \mathbb{R})$, where N is the number of matrices and $\mathcal{X} \subseteq \mathbb{R}^n \otimes \mathbb{R}^p$ is an original matrix input space. $\mathbb{R}^n$ and $\mathbb{R}^p$ are two vector spaces. SMR consists to find a function $f(x)$ that has at most ϵ deviation from the observed y_i for all the training data, and at the same time it is as flat as possible. $f(x)$ will take the form:

$$f(x) = tr(\mathbf{W}'\varphi(\mathbf{X}_j)) + b \tag{9}$$

where $\mathbf{W} \in \mathcal{X}$, $tr(.)$ is the trace operator, and $b \in \mathbb{R}$.

This consists in solving the following convex optimization primal problem:

$$\begin{aligned} \underset{\mathbf{W}}{\text{minimize}} \quad & \frac{1}{2}tr(\mathbf{W}'\mathbf{W}) + C \sum_{j=1}^{N} \left(\xi_j^+ + \xi_j^-\right), \\ \text{subject to} \quad & y_j - \left(tr(\mathbf{W}'\varphi(\mathbf{X}_j)) + b\right) \leq \epsilon + \xi_j^+, \\ & \left(tr(\mathbf{W}'\varphi(\mathbf{X}_j)) + b\right) - y_j \leq \epsilon - \xi_j^-, \\ & \xi_j^+, \xi_j^- \geq 0, \quad j = 1, 2, \ldots, N. \end{aligned} \tag{10}$$

where $C > 0$ is a pre-specified value and determines the trade-off between the flatness of f and the amount up to which deviations larger than ϵ are tolerated, and ξ_j^+, ξ_j^- are slack variables representing upper and lower constraints on the outputs of the system.

3.2 Lagrange Dual Problems

Using similar arguments to the previous section, Slater's conditions are met and thus strong duality holds. It follows that the duality gap is zero and the optimal values of the primal and dual problems are equal. Consequently, it ensures that the original problem (primal) can be solved through the Lagrange dual problem, which is usually easier to solve than the primal. To solve the primal problem (10), we construct the Lagrangian by using Lagrange multipliers. The Lagrangian is

$$
L_p = \frac{1}{2} tr(\mathbf{W}'\mathbf{W}) + C \sum_{j=1}^{N} \left(\xi_j^+ + \xi_j^- \right) - \sum_{j=1}^{N} \alpha_i^+ \left(\epsilon + \xi_j^+ - y_j + tr(\mathbf{W}'\varphi(\mathbf{X}_j)) + b \right)
$$

$$
- \sum_{j=1}^{N} \alpha_i^- \left(\epsilon - \xi_j^- + y_j - tr(\mathbf{W}'\varphi(\mathbf{X}_j)) - b \right) - \sum_{j=1}^{N} \left(\eta_j^+ \xi_j^+ + \eta_j^- \xi_j^- \right)
$$

$$
\tag{11}
$$

It follows from the saddle point condition that the partial derivatives of L_p with respect to the primal variables $(\mathbf{W}, b, \xi_j^+, \xi_j^-)$ have to vanish for optimality.

$$
\frac{\partial L_p}{\partial \mathbf{W}} = 0 \Longrightarrow \mathbf{W} = \sum_{j=1}^{N} (\alpha_j^+ - \alpha_j^-) \varphi(\mathbf{X}_j), \tag{12}
$$

$$
\frac{\partial L_p}{\partial b} = 0 \Longrightarrow \sum_{j=1}^{N} (\alpha_j^+ - \alpha_j^-) = 0, \tag{13}
$$

$$
\frac{\partial L_p}{\partial \xi_j^+} = 0 \Longrightarrow C - \alpha_j^+ - \eta_j^+ = 0. \tag{14}
$$

$$
\frac{\partial L_p}{\partial \xi_j^-} = 0 \Longrightarrow C - \alpha_j^- - \eta_j^- = 0. \tag{15}
$$

Substituting these in the Lagrangian (11) yields the dual problem:

$$
\underset{\alpha^+, \alpha_j^-}{\text{Maximize}} \quad -\frac{1}{2} \sum_{i,j=1}^{N} (\alpha_i^+ - \alpha_i^-)(\alpha_j^+ - \alpha_j^-) tr(\varphi(\mathbf{X}_i)'\varphi(\mathbf{X}_j))
$$

$$
+ \sum_{j=1}^{N} \alpha_j^+ (y_j - \epsilon) - \sum_{j=1}^{N} \alpha_j^- (y_j + \epsilon), \tag{16}
$$

$$
\text{subject to} \quad 0 \leq \alpha_j^+, \alpha_j^- \leq C \quad \forall j,
$$

$$
\sum_{j=1}^{N} (\alpha_j^+ - \alpha_j^-) = 0.
$$

The KKT conditions are

(i) Primal feasibility:

$$y_j - tr(\mathbf{W}'\varphi(\mathbf{X}_j)) - b - \epsilon - \xi_j^+ \leq 0, \tag{17}$$

$$tr(\mathbf{W}'\varphi(\mathbf{X}_j)) + b - y_j - \epsilon - \xi_j^- \leq 0, \tag{18}$$

$$\xi_j^+ \geq 0 \quad \text{and} \quad \xi_j^- \geq 0. \tag{19}$$

(ii) Dual feasibility:

$$\alpha_j^+ \geq 0 \quad \text{and} \quad \alpha_j^- \geq 0 \tag{20}$$

$$\eta_j^+ \geq 0 \quad \text{and} \quad \eta_j^- \geq 0. \tag{21}$$

(iii) Complementary slackness:

$$\alpha_j^+ \left(y_j - tr(\mathbf{W}'\varphi(\mathbf{X}_j)) - b - \epsilon - \xi_j^+ \right) = 0, \tag{22}$$

$$\alpha_j^- \left(tr(\mathbf{W}'\varphi(\mathbf{X}_j)) + b - y_j - \epsilon - \xi_j^- \right) = 0, \tag{23}$$

$$(C - \alpha_j^+)\xi_j^+ = 0 \quad \text{and} \quad (C - \alpha_j^-)\xi_j^- = 0. \tag{24}$$

From the KKT conditions, some important conclusions can be made:

- Samples $(\mathbf{X}_j, y_j)$ with $\alpha_j^+ = C$ or $\alpha_j^- = C$ lie outside the ϵ-insensitive tube around $f(x)$.
- $\alpha_j^+ \alpha_j^- = 0$ meaning that there can never be a set of dual variables α_j^+, α_j^- which are both nonzero.
- For $\alpha_j^+, \alpha_j^- \in (0, C)$, the slack variables ξ_j^+ or ξ_j^- would correspondingly be zero.

By looking at the complementary slackness conditions (Eqs. (22) and (23)), the matrices $\mathbf{X}_j$ with $\alpha_j^+ = 0$ or $\alpha_j^- = 0$ are not needed. Next, the matrices $\mathbf{X}_j$, for which the Lagrange multipliers are nonzero, are needed to determine $\mathbf{W}$ and are called **support matrices** (SM).

To compute the bias term b, one has to choose a support matrix $\mathbf{X}_s$. Then

$$b = y_s - \epsilon - tr(\mathbf{W}'\varphi(\mathbf{X}_s)), \quad \alpha_s^+ \in (0, C), \tag{25}$$

$$b = y_s + \epsilon - tr(\mathbf{W}'\varphi(\mathbf{X}_s)), \quad \alpha_s^- \in (0, C). \tag{26}$$

So for a new observation $\mathbf{Z}$, its prediction is obtained by using the regression function given by

$$f(\mathbf{Z}) = \sum_{SMs} (\alpha_j^+ - \alpha_j^-) tr \left(\varphi(\mathbf{X}_i)' \varphi(\mathbf{X}_j) \right) + b. \tag{27}$$

4 Non-spherical Decision Boundaries

It is not very clear in the SMM methods proposed by Luo et al. (2015) and Zheng et al. (2017), the kernel function used. We propose one alternative of choice for the kernel function. The optimization problem (16) only involves the patterns $\varphi(\mathbf{X})$ through the computation of inner products in feature space. There is no need to compute the features $\varphi(\mathbf{X})$ when one knows how to compute the dot products directly. Instead of actually mapping each instance to a higher dimensional space using a mapping function φ, Boser et al. (1992) propose to directly choose a kernel function $K(\mathbf{X}, \mathbf{Y})$ that represents an inner product $tr(\varphi(\mathbf{X})'\varphi(\mathbf{Y}))$ in some unspecified high dimensional space.

The key idea of the kernel technique, or the so-called kernel trick, is to invert the chain of arguments, i.e., choose a kernel K rather than a mapping before applying a learning algorithm. It is clear that not any symmetric function K can serve as a kernel. The necessary and sufficient conditions of $K : \mathcal{X} \times \mathcal{X} \to \mathbb{R}$ to be a kernel are given by Mercer's theorem.

Theorem 4.1 (Mercer's Theorem) *Suppose K is a symmetric function such that the integral operator*

$$(T_K f)(.) = \int_{\mathcal{X}} K(., \mathbf{x}) f(\mathbf{x}) \, d\mathbf{x}$$

is positive semidefinite, that is,

$$\int_{\mathcal{X}} \int_{\mathcal{X}} K(\mathbf{y}, \mathbf{x}) f(\mathbf{x}) f(\mathbf{y}) \, d\mathbf{x} \, d\mathbf{y} \geq 0,$$

for all integrable functions f. Let $\psi_i \in L_2(\mathcal{X})$ be an eigenfunction of T_k associated with the eigenvalue $\lambda_i \geq 0$ and normalized such that $||\psi_i||_2 = \int_{\mathcal{X}} \psi_i^2(\mathbf{x}) \, d\mathbf{x} = 1$, i.e.,

$$\forall x \in \mathcal{X}, \quad \int_{\mathcal{X}} K(\mathbf{x}, \mathbf{y}) \psi_i(\mathbf{y}) \, d\mathbf{y} = \lambda_i \psi_i(\mathbf{x}).$$

Then

1. $\lambda_i \in \ell_1, \quad i \in \mathbb{N}$
2. $\psi_i \in L_\infty(\mathcal{X})$
3. K can be expanded in a uniformly convergent series, i.e.,

$$K(\mathbf{x}, \mathbf{y}) = \sum_{i=1}^{\infty} \lambda_i \psi_i(\mathbf{x}) \psi_i(\mathbf{y})$$

holds for all $\mathbf{x}, \mathbf{y} \in (X)$.

Mercer's theorem not only gives necessary and sufficient conditions for K to be a kernel, but also suggests a constructive way of obtaining features ϕ_i from a given kernel K.

Theorem 4.2 (Mercer Kernels) *The function $K : \mathcal{X} \times \mathcal{X} \to \mathbb{R}$ is a Mercer kernel if, and only if, for each $l \in \mathbb{N}$ and $\mathbf{X} = (\mathbf{x}_1, \dots, \mathbf{x}_l) \in \mathcal{X}^l$, the $l \times l$ matrix $\mathbf{K} = \left(K(\mathbf{x}_i, \mathbf{x}_j) \right)_{i,j=1}^{l}$ is positive semidefinite.*

Theorem 4.3 (Mercer Condition, Mercer 1909) *The symmetry function $K(\mathbf{x}, \mathbf{y})$ is a valid kernel function if and only if: for all function $f \neq 0$ which satisfies the condition of $\int_{\mathcal{X}} f^2(\mathbf{x}) d\mathbf{x} < \infty$, we need to satisfy the condition:*

$$\int_{\mathcal{X}} \int_{\mathcal{X}} K(\mathbf{y}, \mathbf{x}) f(\mathbf{x}) f(\mathbf{y}) \, d\mathbf{x} \, d\mathbf{y} \geq 0. \tag{28}$$

If we can find a Mercer kernel K that computes an inner product in the feature space $\mathcal{F}$ we are interested in, we can use the kernel evaluations $K(\mathbf{X}, \mathbf{Y})$ to replace the inner products $tr \left(\varphi(\mathbf{X})' \varphi(\mathbf{Y}) \right)$ in the LS-SMM algorithm. Note that obtaining a Mercer kernel in a matrix space is not as easy as in a vector space.

4.1 Wavelet Kernels for Support Matrix Machines

We will construct wavelet kernels that are admissible support matrix kernels, i.e., satisfy Mercer conditions. The support matrix kernel function can be described as the inner product of two matrices, $K(\mathbf{X}, \mathbf{Y}) = tr \left(\varphi(\mathbf{X})' \varphi(\mathbf{Y}) \right)$. Instead of working with matrices, we will use vectors instead. The support vector kernel function can also be described as translation-invariant kernels such as $K(\mathbf{x}, \mathbf{y}) = K(\mathbf{x} - \mathbf{y})$ (Burges 1999). A function is an admissible support vector kernel function if it satisfies the condition of Mercer (Theorem 4.3). However, it is very challenging to decompose translation-invariant functions as the product of two functions and then prove that they satisfy Mercer condition. So the next result gives us necessary and sufficient conditions for translation-invariant kernels to be admissible support vector kernels.

Theorem 4.4 (Smola et al. 1998; Burges 1999) *The translation-invariant kernel function is an admissible support vector kernel function if and only if the Fourier transform of $k(x)$ satisfies*

$$F[k(\omega)] = (2\pi)^{-\frac{m}{2}} \int_{\mathcal{X}^m} \exp\left(-j(\omega x) \right) k(x) dx \geq 0. \tag{29}$$

The goal of wavelet analysis is to approximate a function by a family of functions generated by dilations and translations of a function $\Psi(x)$, called the base wavelet or mother wavelet. The wavelet function group is defined as:

$$\Psi_{a,c}(x) = (a)^{-1/2}\Psi\left(\frac{x-c}{a}\right) \tag{30}$$

where $a \geq 0$ and $c, x \in \mathbb{R}$. a is the dilation factor and c is a translation factor.

If the wavelet function of one dimension is $\Psi(x)$, using tensor theory (Zhang and Benveniste 1992), the multidimensional wavelet function can be defined as:

$$\Psi_m(\mathbf{X}) = \prod_{i=1}^{n}\prod_{j=1}^{p}\Psi(x_{ij}) \tag{31}$$

where $\mathbf{X} = \{x_{ij}\}_{i,j} \in \mathbb{R}^n \otimes \mathbb{R}^p$ is a matrix with entries $x_{i,j}$.

We can build the admissible kernel function for matrix as

Theorem 4.5 *Let Ψ be a base wavelet or mother wavelet, let $a \geq 0$ be the dilation, and $c \in \mathbb{R}$ be the translation. If $\mathbf{X}, \mathbf{Y} \in \mathbb{R}^n \otimes \mathbb{R}^p$ are two matrices, then the wavelet kernels for matrices are*

$$K(\mathbf{X}, \mathbf{Y}) = \prod_{i=1}^{n}\prod_{j=1}^{p}\Psi\left(\frac{x_{ij}-c}{a}\right)\Psi\left(\frac{y_{ij}-c}{a}\right). \tag{32}$$

Proof We just need to show that the wavelet kernels for matrices satisfy the condition of Mercer, i.e., are admissible support matrix kernels.

First, let $\text{vec}(\mathbf{X}) = [x_{11}, \ldots, x_{np}]'$ and $\text{vec}(\mathbf{Y}) = [y_{11}, \ldots, y_{np}]'$, and let set $\mathbf{x} = [x_1, \ldots, x_N]' = \text{vec}(\mathbf{X})$ and $\mathbf{y} = [y_1, \ldots, y_N]' = \text{vec}(\mathbf{Y})$, where $N = n \times p$, then

$$K(\mathbf{X}, \mathbf{Y}) = \prod_{i=1}^{n}\prod_{j=1}^{p}\Psi\left(\frac{x_{ij}-c}{a}\right)\Psi\left(\frac{y_{ij}-c}{a}\right)$$

$$= \prod_{i=1}^{N}\Psi\left(\frac{x_i-c}{a}\right)\Psi\left(\frac{y_i-c}{a}\right)$$

$$= K(\mathbf{x}, \mathbf{y}).$$

Now, $\forall f \in L_2(\mathbb{R}^N)$,

$$\iint_{\mathbb{R}^N \otimes \mathbb{R}^N} K(\mathbf{x}, \mathbf{y}) f(\mathbf{x}) f(\mathbf{y}) \, d\mathbf{x} \, d\mathbf{y} = \int_{\mathbb{R}^N} \prod_{i=1}^{N} \Psi\left(\frac{x_i - c}{a}\right) d\mathbf{x} \int_{\mathbb{R}^N} \prod_{i=1}^{N} \Psi\left(\frac{y_i - c}{a}\right) d\mathbf{y}$$

$$= \left(\int_{\mathbb{R}^N} \prod_{i=1}^{N} \Psi\left(\frac{x_i - c}{a}\right) d\mathbf{x}\right)^2 \geq 0.$$

$\square$

Therefore, $K(\mathbf{X}, \mathbf{Y})$ satisfies Mercer condition and is admissible support matrix kernel. Consequently, it follows that we can build translation-invariant kernels as follows:

$$K(\mathbf{X}, \mathbf{Y}) = \prod_{i=1}^{n} \prod_{j=1}^{p} \Psi\left(\frac{x_{ij} - y_{ij}}{a}\right). \tag{33}$$

A necessary and sufficient condition for translation-invariant kernels to be admissible is to satisfy the condition of Mercer (Theorem 4.4).

Mexican Hat Wavelet Kernel for Support Matrix

Now, we give an existing wavelet kernel function, the Mexican hat wavelet or Sombrero wavelet, that can be used to construct translation-invariant wavelet kernels. Note that the Mexican hat wavelet is sometimes called Marr wavelet and its mother wavelet is

$$\Psi(x) = (1 - x^2) \exp\left(-\frac{x^2}{2}\right). \tag{34}$$

Figure 1 represents a 2-D plot of the Mexican Hat wavelet kernel function.

Theorem 4.6 *Let $a \geq 0$, if $\mathbf{X}, \mathbf{Y} \in \mathbb{R}^n \otimes \mathbb{R}^p$ are two matrices, then the Mexican Hat wavelet kernel function for matrices is defined as:*

$$K(\mathbf{X}, \mathbf{Y}) = \prod_{i=1}^{n} \prod_{j=1}^{p} \left(1 - \left(\frac{x_{ij} - y_{ij}}{a}\right)^2 \exp\left(-\frac{(x_{ij} - y_{ij})^2}{2a^2}\right)\right), \tag{35}$$

and is an admissible support matrix kernel.

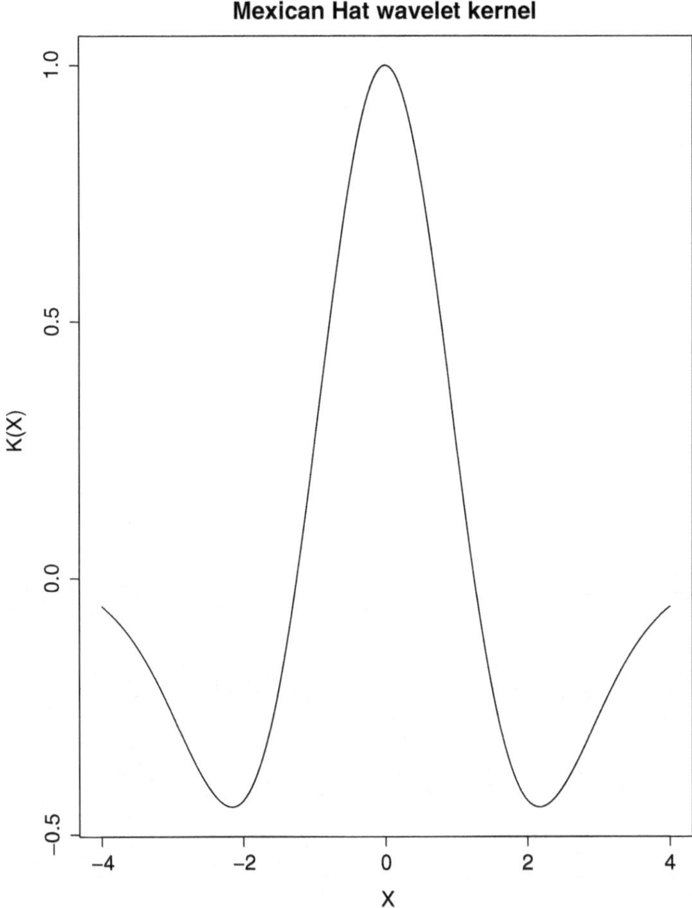

Fig. 1 Mexican Hat wavelet kernel function

Proof According to Theorem 4.4, one just needs to show that the Fourier transform of Mexican hat wavelet is nonnegative. Let $\mathbf{x} = [x_1, \ldots, x_N]' = \text{vec}(\mathbf{X})$ with $N = n \times p$, then it follows that

$$K(x) = \prod_{i=1}^{N} \Psi\left(\frac{x_i}{a}\right)$$

$$= \prod_{i=1}^{N} \left(1 - \left(\frac{x_i}{a}\right)^2\right) \exp\left(-\frac{1}{2}\left(\frac{x_i}{a}\right)^2\right).$$

The integral term

$$
I = \int_{\mathbb{R}^N} \exp(-j(\boldsymbol{\omega}\mathbf{x})) K(x)\,d\mathbf{x}
$$

$$
= \int_{\mathbb{R}^N} \exp(-j(\boldsymbol{\omega}\mathbf{x})) \left(\prod_{i=1}^{N} \left(1 - \left(\frac{x_i}{a}\right)^2 \right) \exp\left(-\frac{1}{2}\left(\frac{x_i}{a}\right)^2 \right) \right) d\mathbf{x}
$$

$$
= \prod_{i=1}^{N} \int_{\infty}^{\infty} \exp\left(-ja\omega_i \frac{x_i}{a} \right) \left(\left(1 - \frac{x_i^2}{a^2} \right) \exp\left(-\frac{x_i^2}{2a^2} \right) \right) dx_i
$$

$$
= a^N \prod_{i=1}^{N} \int_{-\infty}^{\infty} \exp\left(-ja\omega_i \frac{x_i}{a} \right) \left(\left(1 - \frac{x_i^2}{a^2} \right) \exp\left(-\frac{x_i^2}{2a^2} \right) \right) d\left(\frac{x_i}{a}\right)
$$

$$
= a^N \prod_{i=1}^{N} \int_{-\infty}^{\infty} \left(1 - \frac{x_i^2}{a^2} \right) \exp\left(-\frac{x_i^2}{2a^2} - ja\omega_i \frac{x_i}{a} \right) d\left(\frac{x_i}{a}\right)
$$

$$
= a^N \prod_{i=1}^{N} \int_{-\infty}^{\infty} \exp\left(-\frac{x_i^2}{2a^2} - ja\omega_i \frac{x_i}{a} \right) d\left(\frac{x_i}{a}\right)
$$

$$
- \int_{-\infty}^{\infty} \frac{x_i^2}{2a^2} \exp\left(-\frac{x_i^2}{2a^2} - ja\omega_i \frac{x_i}{a} \right) d\left(\frac{x_i}{a}\right)
$$

$$
= a^N \prod_{i=1}^{N} \int_{-\infty}^{\infty} \exp\left(-\frac{1}{2}\left(x_i^2 + 2ja\omega_i x_i \right) \right) dx_i
$$

$$
- \int_{-\infty}^{\infty} x_i^2 \exp\left(-\frac{1}{2}\left(x_i^2 + 2ja\omega_i x_i \right) \right) dx_i
$$

$$
= a^N \prod_{i=1}^{N} \exp\left(-\frac{1}{2} j^2 a^2 \omega_i^2 \right) \int_{-\infty}^{\infty} \exp\left(-\frac{1}{2}(x_i + ja\omega_i)^2 \right) dx_i
$$

$$
- \int_{-\infty}^{\infty} x_i^2 \exp\left(-\frac{1}{2}(x_i + ja\omega_i)^2 \right) dx_i
$$

$$
= a^N \exp\left(-\frac{1}{2} j^2 a^2 \sum_{i=1}^{N} \omega_i^2 \right) \prod_{i=1}^{N} \{G_1(\boldsymbol{\omega}) - G_2(\boldsymbol{\omega})\}.
$$

where $G_1(\boldsymbol{\omega}) = \int_{-\infty}^{\infty} \exp\left(-\frac{1}{2}(x_i + ja\omega_i)^2 \right) dx_i$ and $G_2(\boldsymbol{\omega}) = \int_{-\infty}^{\infty} x_i^2 \exp\left(-\frac{1}{2}(x_i + ja\omega_i)^2 \right) dx_i$.

Using partial integration, we get

$$G_1(\omega) = (2\pi)^{1/2} \quad \text{and} \quad G_2(\omega) = (2\pi)^{1/2}(1 - a^2\omega^2). \tag{36}$$

Then

$$\prod_{i=1}^{N} \{G_1(\omega) - G_2(\omega)\} = (2\pi)^{N/2} a^{2N} \omega_i^{2N}. \tag{37}$$

and so

$$I = (2\pi)^{N/2} a^{3N} \exp\left(-\frac{1}{2} j^2 a^2 \sum_{i=1}^{N} \omega_i^2\right) \prod_{i=1}^{N} \omega_i^2. \tag{38}$$

It follows therefore that

$$F(K(\omega)) = (2\pi)^{N/2} I = (2\pi)^N a^{3N} \exp\left(-\frac{1}{2} j^2 a^2 \sum_{i=1}^{N} \omega_i^2\right) \prod_{i=1}^{N} \omega_i^2. \tag{39}$$

and $F(K(\omega)) \geq 0$ since $a \geq 0$ and the proof is completed. $\qquad\square$

Morlet Wavelet Kernel for Support Matrix

Another wavelet kernel function that can be used is Morlet wavelet kernel function. Its mother wavelet is defined as:

$$\Psi(\mathbf{x}) = \cos(\omega_0 \mathbf{x}) \exp\left(-\frac{x^2}{2}\right). \tag{40}$$

Figure 2 represents a 2-D plot of the Morlet wavelet kernel function. Similarly, we define the Morlet wavelet kernel function as follows.

Theorem 4.7 *Let $a \geq 0$, if $\mathbf{X}, \mathbf{Y} \in \mathbb{R}^n \otimes \mathbb{R}^p$ are two matrices, then the Morlet wavelet kernel function for matrices is defined as:*

$$K(\mathbf{X}, \mathbf{Y}) = \prod_{i=1}^{n} \prod_{j=1}^{p} \Psi\left(\frac{x_{ij} - y_{ij}}{a}\right) = \prod_{i=1}^{n} \prod_{j=1}^{p} \cos\left(\omega\left(\frac{x_{ij} - y_{ij}}{a}\right)\right) \exp\left(-\frac{(x_{ij} - y_{ij})^2}{2a^2}\right). \tag{41}$$

and is an admissible support matrix kernel function.

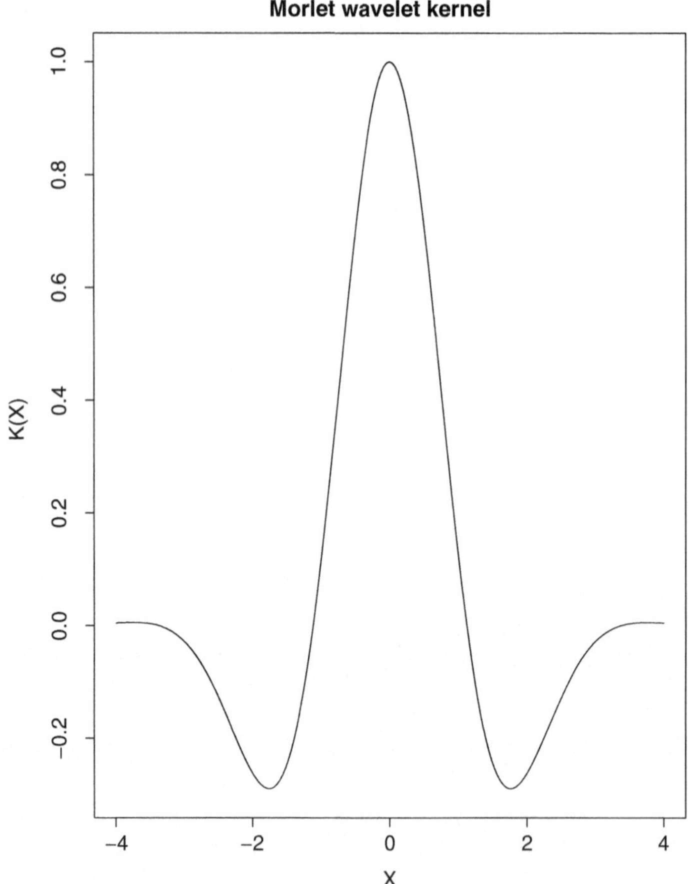

Fig. 2 Morlet wavelet kernel function

Proof Using Theorem 4.4, we need to show that the Fourier transform of Morlet wavelet is nonnegative. Let $\mathbf{x} = [x_1, \ldots, x_N]' = \text{vec}(\mathbf{X})$ with $N = n \times p$, then it follows that

$$K(x) = \prod_{i=1}^{N} \Psi\left(\frac{x_i}{a}\right)$$

$$= \prod_{i=1}^{N} \left(\cos\left(\omega_0 \frac{x_i}{a}\right)\right) \exp\left(-\frac{1}{2}\left(\frac{x_i}{a}\right)^2\right).$$

The integral term

$$J = \int_{\mathbb{R}^N} \exp(-j(\boldsymbol{\omega}\mathbf{x})) K(x) d\mathbf{x}$$

$$= \int_{\mathbb{R}^N} \exp(-j(\boldsymbol{\omega}\mathbf{x})) \left(\prod_{i=1}^{N} \left(\cos\left(\omega_0 \frac{x_i}{a} \right) \right) \exp\left(-\frac{1}{2} \left(\frac{x_i}{a} \right)^2 \right) \right) d\mathbf{x}$$

$$= a^N \prod_{i=1}^{N} \int_{-\infty}^{\infty} \left(\cos\left(\omega_0 \frac{x_i}{a} \right) \right) \exp\left(-\frac{x_i^2}{2a^2} - ja\omega_i \frac{x_i}{a} \right) d\left(\frac{x_i}{a} \right)$$

$$= a^N \prod_{i=1}^{N} \int_{-\infty}^{\infty} \exp\left(-\frac{x_i^2}{2a^2} - ja\omega_i \frac{x_i}{a} \right)$$

$$\times \frac{1}{2} \left(\exp\left(\omega_0 \frac{x_i}{a} \right) + \exp\left(-\omega_0 \frac{x_i}{a} \right) \right) d\left(\frac{x_i}{a} \right)$$

$$= \left(\frac{a}{2} \right)^N \prod_{i=1}^{N} \int_{-\infty}^{\infty} \exp\left(-\frac{x_i^2}{2a^2} + (j\omega_0 - ja\omega_i) \left(\frac{x_i}{a} \right) \right) dx_i$$

$$+ \int_{-\infty}^{\infty} \exp\left(-\frac{x_i^2}{2a^2} - (j\omega_0 + ja\omega_i) \left(\frac{x_i}{a} \right) \right) dx_i$$

$$= a^N \prod_{i=1}^{N} \frac{(2\pi)^{1/2}}{2} \left\{ \exp\left(-\frac{(\omega_0 + a\omega_i)^2}{2} \right) + \exp\left(-\frac{(\omega_0 - a\omega_i)^2}{2} \right) \right\}.$$

It follows now that

$$F(K(\boldsymbol{\omega})) = (2\pi)^{N/2} J$$

$$= \left(\frac{a}{2} \right)^N \prod_{i=1}^{N} \left\{ \exp\left(-\frac{(\omega_0 + a\omega_i)^2}{2} \right) + \exp\left(-\frac{(\omega_0 - a\omega_i)^2}{2} \right) \right\}. \tag{42}$$

and $F(K(\boldsymbol{\omega})) \geq 0$ since $a \geq 0$ and the proof is completed.

$\square$

The left panel of Fig. 3 shows 3-D plot of the Mexican Hat kernel function while the right panel displays a 3-D plot of Morlet kernel. The two plots look distinctly different from each other.

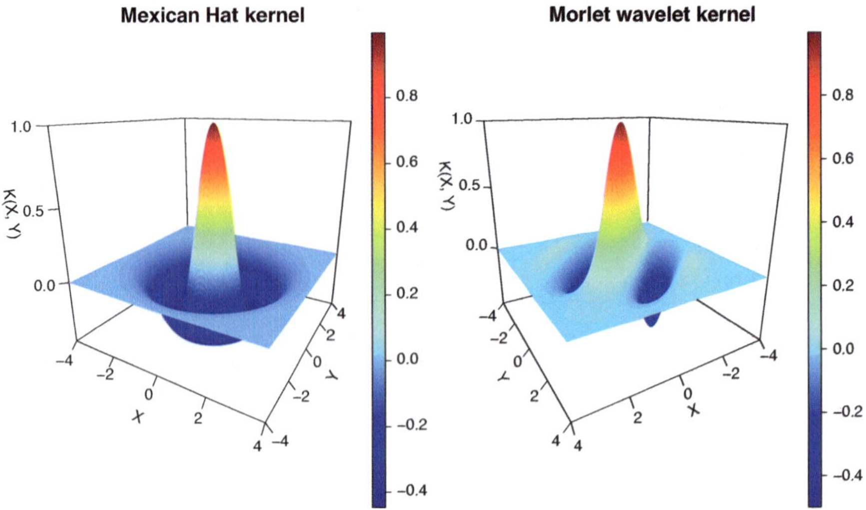

Fig. 3 3-D plot of wavelet kernels. (left) Mexican Hat wavelet kernel, (right) Morlet wavelet kernel

5 Applications

We illustrate the use of the two wavelet kernel functions on two datasets. We apply WSMM to EEG and image classification problems. The EEG alcoholism dataset is concerned with the relationship between genetic predisposition and tendency for alcoholism. The study involved two groups of subjects: an alcoholic and a control group. Each subject was exposed to a stimulus while voltage values were measured from 64 channels of electrodes placed on the subject's scalp for 256 time points. So each subject has measurements of electrical scalp activity, which form a 256×64 matrix. There are 77 subjects from the alcoholic group and 45 subjects from the control group. In our application, we used both 10 subjects from the alcoholic and control groups. The performance comparison was assessed in terms of the classification accuracy. Both the Mexican Hat wavelet kernel function for matrices (Eq. (35)) and Morlet wavelet kernel function for matrices (Eq. (41)) yield an accuracy of 95%.

The second dataset used is the INRIA person dataset. This dataset was proposed to detect whether or not people exist in an image. Each color image is converted into a 160×96 gray level image and the pixel values are used as an input matrix without any advanced feature extraction technique. We use a small subset of the dataset. The training set has 60 positives and 30 negatives for a total of 90 matrices. The test set consists of 55 positives and 25 negatives for a total of 80 matrices. The Mexican Hat wavelet kernel function for matrices (Eq. (35)) gives a classification accuracy of 87.5% while Morlet wavelet kernel function for matrices, (Eq. (41)) yields a classification accuracy of 88.5%.

6 Conclusion

This article proposed some new kernel functions of support matrix machines, Mexican Hat, and Morlet wavelet kernel functions. These kernel functions were used to map matrices from the low dimensional matrix space to some high dimensional space. We establish by proving that these kernels are valid or admissible kernels. The method was successfully applied to EEG and INRIA image classification with good performances.

References

Boser, B.E., Guyon, I., Vapnik, V.N.: A training algorithm for optimal margin classifiers. In: Proceedings of the Fifth Annual Workshop of Computational Learning Theory, vol. 5, pp. 144–152. ACM, Pittsburgh (1992)

Burges, C.J.C.: Geometry and invariance in kernel based methods. In: Scholkopf, B., Burges, C.J.C., Smola, A.J. (eds.) Advances in Kernel Methods–Support Vector Learning, pp. 89–116. MIT, Cambridge (1999)

Cortes, C., Vapnik, V.: Support vector networks. Mach. Learn. **20**, 273–297 (1995)

Luo, L., Xie Y., Zhang, Z., Li, W.-J.: Support matrix machines. In: Proceedings of the 32nd International Conference on Machine Learning (ICML), pp. 928–947 (2015)

Maboudou-Tchao, E.M.: Kernel methods for changes detection in covariance matrices. Commun. Stat. Simul. Comput. (2017). http://dx.doi.org/10.1080/03610918.2017.1322701

Mercer, J.: Functions of positive and negative type and their connection with the theory of integral equations. Philos. Trans. R. Soc. Lond. A **209**, 415–446 (1909)

Shi, W., Zhang, D.: Support matrix machine for large-scale data set. In: International Conference on Information Engineering and Computer Science, 2009. ICIECS 2009. 20, pp. 1191–1199 (2009)

Smola, A.J.: Regression estimation with support vector learning machines. Master's thesis, Technische Universitat Munchen (1996)

Smola, A., Scholkopf, B., Muller, K.-R.: The connection between regularization operators and support vector kernels. Neural Netw. **11**, 637–649 (1998)

Vapnik, V.: The Nature of Statistical Learning Theory. Springer, New York (1995). ISBN 0-387-94559-8

Wu, F., Zhao, Y.: Least squares support vector machine on Morlet wavelet kernel function and its application to nonlinear system identification. Inf. Technol. J. **5**(3), 439–444 (2006)

Xia, W., Fan, L.: Least squares support matrix machines based on bilevel programming. Int. J. Appl. Math. Mach. Learn. **4**(1), 1–18 (2016)

Zhang, Q., Benveniste, A.: Wavelet networks. IEEE Trans. Neural Netw. **3**(6), 889–898 (1992)

Zhang, L., Zhou, W., Jiao, L.: Wavelet support vector machine. IEEE Trans. Syst. Man Cybern. B Cybern. **34**, 34–39 (2004)

Zheng, Q., Zhu, F., Qin, J., Chen, B., Heng, P.H.: Sparse support matrix machine. Pattern Recogn. **76**, 1–12 (2017)

Properties of the Number of Iterations of a Feasible Solutions Algorithm

Sarah A. Janse and Katherine L. Thompson

1 Introduction

Many approaches exist to identify interaction effects in data sets with small to moderate size. Classical statistical methods suggest considering all pairwise combinations of possible explanatory variables in the proposed logistic regression or linear regression model, and selecting a set of variables based on either hypothesis tests or a model selection criterion. Although the theory supporting these techniques is developed, often the data sets of interest have an inordinate number of possible explanatory variables to consider in higher-order interactions using the conventional implementations of classical methods.

For example, genomic data is also unique in its complexity due to intricate dependencies among genes and traits, often in the form of information from external influences or genetic makeup that are unaccounted for during analysis. In particular, interaction effects among genes contribute to epistasis, which is especially difficult to identify in genomic data (Moore and Williams 2009). These interactions have been targets of recent analyses of genomic data although development of methods to identify higher-order interaction effects has been more limited due to computational concerns (Gemperline 1999). For example, even using the second largest

Electronic Supplementary Material The online version of this chapter (https://doi.org/10.1007/978-3-030-11431-2_5) contains supplementary material, which is available to authorized users.

S. A. Janse
Center for Biostatistics, The Ohio State University, Columbus, OH, USA

K. L. Thompson (✉)
Department of Statistics, University of Kentucky, Lexington, KY, USA
e-mail: katherine.thompson@uky.edu

© Springer Nature Switzerland AG 2019

N. Diawara (ed.), *Modern Statistical Methods for Spatial and Multivariate Data*,
STEAM-H: Science, Technology, Engineering, Agriculture, Mathematics & Health,
https://doi.org/10.1007/978-3-030-11431-2_5

supercomputer at IBM (a 262,144 core machine), Goudey et al. performed analyses examining two-way interactions on a large data set of 1.1 million single nucleotide polymorphisms (SNPs) from 2000 samples in less than 10 min, and project that analyzing three-way interactions on the same computer would take approximately 5.8 years (Goudey et al. 2015). Thus, exhaustive searches using anything less than a massive supercomputer are highly impractical for this type of data, especially in the case of higher-order interactions.

In contrast, stochastic search algorithms address the computational concern of the exhaustive search methods by employing some aspect of randomness in order to perform a non-exhaustive search over the possible explanatory variables. However, these methods may not produce the same result every time, and thus may fail to identify the truly optimal model according to the criterion used. In addition, most exhaustive and stochastic searches produce a single "best" model or set of explanatory variables with respect to some hypothesis testing or model selection criterion. In any given data set, there may be another, nearly best model that is more practical than the statistically best model. Only considering the single statistically optimal solution leaves little room to consider more practically meaningful combinations of variables without further experimentation.

Thus, room for improvement exists in implementing methods that are fast and flexible in their ability to detect both main and interaction effects in a model, and reliable in detecting effects that are not only statistically, but also practically, significant. FSA may produce several nearly optimal solutions from different iterations of the algorithm, rather than a single optimal solution. This variability in FSA produces multiple results for consideration to glean practically reasonable conclusions from the data, rather than ending with a single (statistically) optimal solution. In the latter case, one solution may be optimal and practically nonsensical, while another nearly optimal solution exists that is biologically relevant. FSA will show the analyst both solutions for consideration during analysis.

Due to the stochastic nature of FSA, one issue that arises when implementing the algorithm is the choice of number of iterations. Here, each replication is referred to as a random start and begins with an arbitrary model based on the desired mth-order interaction. The number of random starts must be chosen by the user. Thus, we derive a bound on the probability of obtaining the statistically optimal solution in a set number of random starts of FSA that can be used to select this number. This allows users to choose a bound such that they obtain the statistically optimal solution with a desired probability, prior to beginning data analysis.

2 Background

Issues from the complex nature of interaction effects, coupled with the size of data, cause theoretical and computational problems when classical methods are applied using standard implementations. To address these limitations, some recent work has been focused on revisiting versions of the Feasible solutions algorithm

(FSA) first popularized by Doug Hawkins at the University of Minnesota in the early 1990s (Miller 1984; Hawkins and Olive 1999). Several versions of FSA exist (Hawkins 1993, 1994a,b, 1993), but we are focused on the current version of FSA, which is used for subset selection. Common algorithms and methods exist to find the best subset of predictors that adequately explain the response variable. Forward selection, backward selection, and stepwise selection are common automatic variable selection techniques. LASSO and ridge regression are common penalized regression techniques that are used for subset selection with many variables. Exhaustive search checks all possible combinations for the possible model structures. These are currently available in the form of R packages for linear and generalized linear models in leaps (Lumley and Miller 2004) and glmnet (Friedman et al. 2009).

Here, we study the properties of FSA, which is designed to find interactions when the number of predictors is large (Lambert et al. 2018). FSA searches the set of all possible interaction effects to identify those that improve the predictive model for a given response. Issues from the complex nature of interaction effects, coupled with the size of big data, cause theoretical and computational problems when classical methods are applied using standard implementations. One advantage of applying FSA in these cases is that it provides more than one feasible solution, or candidate set of explanatory variables, for a particular data analysis. Providing a set of solutions increases the likelihood of finding practically significant associations rather than solely statistically significant associations.

Specifically, FSA is carried out as follows:

1. Randomly choose m variables from the possible p predictors and compute a specified objective function, i.e., R^2.
2. Consider exchanging one of the m selected explanatory variables from the current model with another explanatory variable in the data set.
3. If an improvement exists, make the exchange that improves the objective function the most.
4. Keep making exchanges until the objective function does not improve. The explanatory variables included in the resulting model are called a feasible solution.
5. Repeat steps (1)–(4) for the number of random starts specified to find additional feasible solutions.

A feasible solution is optimal in that no one exchange of a variable in the model for another outside of the model can improve the selected criterion function. Not only does FSA provide a set of feasible solutions, but it is often more computationally efficient than standard exhaustive approaches due to its stochastic nature. Other advantages of the algorithm include the ability to analyze both linear and logistic regression models, as well as being able to implement several different optimization criteria. In this work, we focus on the required number of iterations, or replications, required by FSA to produce the statistically optimal model.

3 Methods

In FSA, each random start begins with an arbitrary model with a fixed number of
predictors and proceeds by taking steps to better models based on some optimization
criteria, e.g., R^2. The algorithm proceeds until it reaches an optimal model for a
given random start. Thus, each random start, or replication of FSA, will have at least
one step, but often times will have several more. FSA is not guaranteed to identify
the optimal solution, but as the analyst increases the number of random starts, FSA
is more likely to do so. Thus, we need enough random starts to obtain the optimal
solution with some probability. However, the larger the number of random starts,
the longer the time it will take FSA to run. Therefore, it would be highly useful to
have information regarding how many random starts to choose in order to obtain
the optimal solution with some probability while still maintaining computational
efficiency.

As the number of explanatory variables, p, in a data set increases, it is more
difficult to identify the optimal solution and will require more random starts. We
propose choosing the number of random starts as a function of p. As p goes to
infinity, the probability that the optimal solution is identified by FSA is bounded
below. The limit described in Theorem 1 holds for FSA in the case of considering
m-way interactions.

Theorem 1 *In the case of using FSA to find a statistically significant m-way
interaction in a predictive model, as the number of potential explanatory variables,
p, goes to infinity, a lower bound on the probability of identifying the statistically
optimal model in cp random starts, where $0 < c < 1$, is $1 - e^{-cm^2}$.*

Lemma

$$\lim_{x \to \infty} \left[1 + \frac{k}{x} \right]^{tx} = e^{tk}$$

Proof of Theorem 1 Let p be the number of possible explanatory variables we are
choosing from, c be a constant such that $0 < c < 1$, and cp be the number of
random starts. Then there are $\binom{p-m}{m}$ pairs of variables out of the total $\binom{p}{m}$ possible
pairs that do not contain any of the variables in the optimal solution, consisting of
m variables. Note that, if you randomly start with $m - 1$ out of the m variables in
the statistically optimal solution, you are guaranteed to obtain the optimal solution.
Then, the probability of not identifying the optimal solution in the first step of a
given random start is

$$\frac{\binom{p-m}{m}}{\binom{p}{m}} \tag{1}$$

and so the probability of obtaining the optimal solution in the first step of a given random start is

$$1 - \frac{\binom{p-m}{m}}{\binom{p}{m}}. \tag{2}$$

For a given random start, FSA completes at least one step, and often more than one step, before reaching a feasible solution. Since we are only considering finding the statistically optimal solution after the first step and not considering the cases where we could find the optimal solution in later steps, Eq. (2) will be a lower bound on the probability of identifying the statistically optimal solution in a single random start. So, the probability of obtaining the statistically optimal solution in at least one of the cp random starts is greater than $1 - \left[\frac{\binom{p-m}{m}}{\binom{p}{m}}\right]^{cp}$, where $\left[\frac{\binom{p-m}{m}}{\binom{p}{m}}\right]^{cp}$ is the probability that none of the random starts identify the optimal solution in the first step of FSA. So we consider

$$\lim_{p\to\infty}\left(1 - \left[\frac{\binom{p-m}{m}}{\binom{p}{m}}\right]^{cp}\right)$$

$$= \lim_{p\to\infty} 1 - \lim_{p\to\infty}\left[\frac{\binom{p-m}{m}}{\binom{p}{m}}\right]^{cp}$$

$$= 1 - \lim_{p\to\infty}\left[\frac{(p-m)!}{m!(p-2m)!}\frac{m!(p-m)!}{p!}\right]^{cp}$$

$$= 1 - \lim_{p\to\infty}\left[\frac{(p-m)!(p-m)!}{p!(p-2m)!}\right]^{cp}$$

$$= 1 - \lim_{p\to\infty}\left[\frac{(p-m)!}{(p-2m)!p(p-1)\cdots(p-m+1)}\right]^{cp}$$

$$= 1 - \lim_{p\to\infty}\left[\frac{(p-m)(p-m-1)\cdots(p-2m+1)}{p(p-1)\cdots(p-m+1)}\right]^{cp}.$$

Notice that both the numerator and denominator in the limit statement contain m quantities. Thus we can write the last line above as

$$= 1 - \lim_{p\to\infty}\left[\frac{p-m}{p}\right]^{cp}\left[\frac{p-m-1}{p-1}\right]^{cp}\cdots\left[\frac{p-2m+1}{p-m+1}\right]^{cp}$$

$$= 1 - \lim_{p\to\infty}\left[\frac{p-m}{p}\right]^{cp}\lim_{p\to\infty}\left[\frac{p-m-1}{p-1}\right]^{cp}\cdots\lim_{p\to\infty}\left[\frac{p-2m+1}{p-m+1}\right]^{cp}$$

$$= 1 - \lim_{p\to\infty}\left[1-\frac{m}{p}\right]^{cp}\lim_{p\to\infty}\left[1-\frac{m}{p-1}\right]^{cp}\cdots\lim_{p\to\infty}\left[1-\frac{m}{p-m+1}\right]^{cp}.$$

Then we have

$$\lim_{p \to \infty} \left[1 - \frac{m}{p} \right]^{cp} = e^{-cm}$$

by the lemma with $t = c$ and $k = -m$. Next,

$$\lim_{p \to \infty} \left[1 - \frac{m}{p-1} \right]^{cp}$$

$$= \lim_{p \to \infty} \left[1 - \frac{m}{p-1} \right]^{c(p-1)} \left[1 - \frac{m}{p-1} \right]^{c}.$$

Since $\lim_{p \to \infty} \left[1 - \frac{m}{p-1} \right]^{c(p-1)} = e^{-cm}$ by the lemma with $t = c$ and $k = -m$

and $\lim_{p \to \infty} \left[1 - \frac{m}{p-1} \right]^{c} = 1$, we have

$$\lim_{p \to \infty} \left[1 - \frac{m}{p-1} \right]^{cp} = e^{-cm}.$$

Next,

$$\lim_{p \to \infty} \left[1 - \frac{m}{p-m+1} \right]^{cp}$$

$$= \lim_{p \to \infty} \left[1 - \frac{m}{p-m+1} \right]^{c(p-m+1)} \left[1 - \frac{m}{p-m+1} \right]^{c(m-1)}.$$

Since $\lim_{p \to \infty} \left[1 - \frac{m}{p-m+1} \right]^{c(p-m+1)} = e^{-cm}$ by the lemma with $t = c$ and

$k = -m$ and $\lim_{p \to \infty} \left[1 - \frac{m}{p-m+1} \right]^{c(m-1)} = 1$, we have

$$\lim_{p \to \infty} \left[1 - \frac{m}{p-m+1} \right]^{cp} = e^{-cm}.$$

So,

$$1 - \lim_{p \to \infty} \left[\frac{\binom{p-m}{m}}{\binom{p}{m}} \right]^{cp} = 1 - e^{-cm} \times e^{-cm} \times \cdots e^{-cm} \, (m \, \text{times})$$

$$= 1 - e^{-c(m^2)}.$$

Fig. 1 In this plot, the dots show the exact value of the lower bound for varied values of c, and the lines show the asymptotic lower bound on the probability of getting the statistically optimal solution with $m = 2$. The lower bound is attained very quickly and the probability of identifying the statistically optimal solution increases as the number of random starts increases, as expected

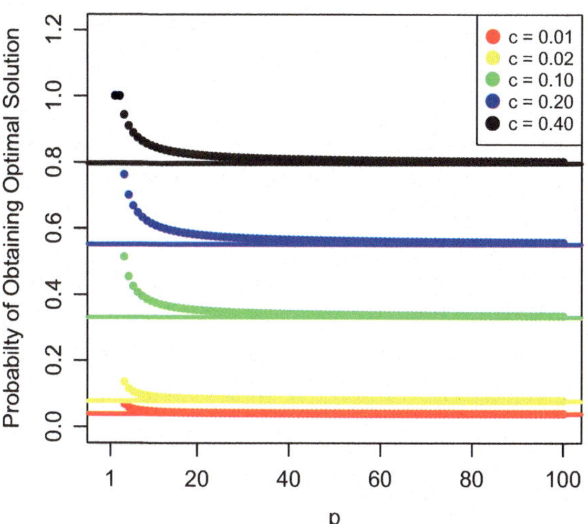

Thus,

P(Obtaining the statistically optimal model in cp random starts using FSA)

$$\geq 1 - \left[\frac{\binom{p-m}{m}}{\binom{p}{m}} \right]^{cp},$$

and note that:

$$\lim_{p \to \infty} \left(1 - \left[\frac{\binom{p-m}{m}}{\binom{p}{m}} \right]^{cp} \right) = 1 - e^{-cm^2}.$$

Figure 1 shows how the calculated probability of obtaining the optimal solution approaches the lower bound derived above for 5 values of c with $m = 2$. It can be seen that the lower bound is attained very quickly and thus is appropriate when considering data sets with a large number of explanatory variables, p. It is also clear that the probability of obtaining the statistically optimal solution increases as the number of starts increases, as is expected.

4 Results

Simulation studies were performed for both quantitative and binary response variables to examine the outcomes of utilizing the lower bound derived above. These simulations were followed by a real data analysis to demonstrate the use of the lower bound in practice.

4.1 Simulations

Quantitative trait data were simulated as the sum of two covariates and their interaction under the typical regression model for values of p of 50, 100, 1000, and 2500. One hundred data sets were simulated for each value of p. Binary trait data were simulated in an analogous manner. Simulations parameters are as follows:

- Quantitative response variable (lmFSA)

  - $X_{ij} \sim U(0, 1)$ for $i = 1, \ldots, n$ and $j = 1, .., p$
  - $Y_i = 5 + X_{i1} + X_{i2} + 2X_{i1}X_{i2} + \epsilon_i$ where $\epsilon_i \sim N(0, 1)$

- Binary response variable (glmFSA)

  - $X_{ij} \sim U(0, 1)$ for $i = 1, \ldots, n$ and $j = 1, \ldots, p$
  - $\pi_i = \dfrac{e^{X_{i1}+X_{i2}+2X_{i1}X_{i2}}}{(1+e^{X_{i1}+X_{i2}+2X_{i1}X_{i2}})}$
  - $Y_i = \begin{cases} 1 & \text{with probability } \pi_i \\ 0 & \text{with probability } 1 - \pi_i \end{cases}$

FSA was used to provide a set of feasible solutions for every simulated data set via the implementation in Lambert et al. (2018). Exhaustive search was then performed to find the statistically optimal solution using R^2 and AIC as the criteria functions for the quantitative and binary response variables, respectively. The numbers of random starts chosen for FSA were values of c including 0.01, 0.02, 0.1, 0.2, and 0.4 with each value of p. Then, for each simulation setting, the percentage of simulated data sets producing the statistically optimal solution using FSA was calculated. These percentages, along with the lower bound from Sect. 3, are plotted in Fig. 2.

Figure 2 shows the results from 100 simulated data sets for both methods in FSA for four values of p and five values of c. Note that the asymptotic lower bound proposed here depends only on m and c. The red dots represent these lower bounds for each value of c. The yellow diamonds, blue dots, green squares, and black triangles represent the percentage of 100 simulations with $p = 50$, 100, 1000, and 2500, respectively, when $m = 2$ [where FSA was able to identify the statistically optimal solution]. It is clear from Fig. 2 that the lower bound is often much lower than the observed probability and is thus very conservative, but does provide a good guidance as to the number of random starts needed to produce at least one feasible solution containing the statistically optimal solution.

4.2 Real Data Example

Data were collected in a genome-wide association study using 288 outbred mice in a study that aimed to identify, or map, locations along the genome called SNPs that influence HDL cholesterol, systolic blood pressure, triglyceride levels, glucose, or

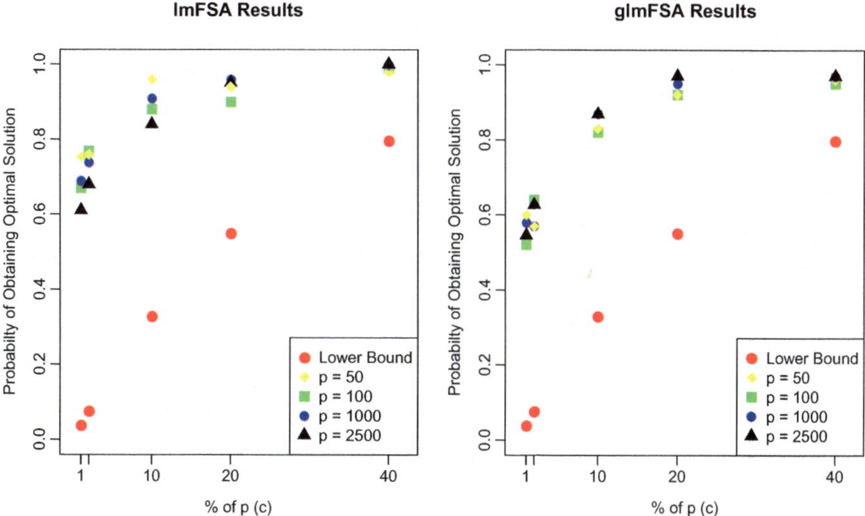

Fig. 2 Simulation results for the probability of getting the optimal solution in 100 simulations with a quantitative response variable (left) and a binary response variable (right): for each value of c, the lower bounds are represented by the red dots and the probability of identifying the statistically optimal solution for the four values of p is represented by yellow diamonds ($p = 50$), blue dots ($p = 100$), green squares ($p = 1000$), and black triangles ($p = 2500$). Both the left and right plots show that the lower bound is valid for all values of p in the simulation study

urinary albumin-to-creatinine ratios (Zhang et al. 2012). Our goal was to determine if SNPs or interactions of SNPs were associated with HDL levels. Information from 3045 SNPs on chromosome 11 were analyzed for this real data analysis.

Using the lower bound in Theorem 1, if we want the probability of obtaining the statistically optimal solution including a 2-way interaction to be at least 95%, then we need to solve the following equation for c:

$$1 - e^{-c2^2} = 0.95$$

$$\iff c = 0.7489331$$

Since we have $p = 3045$, the number of random starts we need is $0.75(3045) = 2283.75$, or 2284 random starts.

The exhaustive search of the 3045 SNPs on chromosome 11 took approximately 11 h in total on a large cluster without parallelization. Using this method, we found that the statistically optimal solution includes an interaction between mb2863979 and mb87344525 and corresponds to a value of $R^2 = 0.1256$, which can be found in Table 1. FSA took approximately half an hour to perform 2284 random starts when parallelized on a larger cluster, using 16 cores. There were a total of 33 feasible solutions identified through FSA, including the statistically optimal solution, which is presented in bold in the subset of FSA results in Table 2. (Full FSA results can

Table 1 The exhaustive search produced the single statistically optimal solution with $R^2 = 0.1256308$ (column 3)

Variable 1	Variable 2	R^2
mb2863979	mb87344525	0.1256308

Columns 1 and 2 show the SNPs that were identified in this model

Table 2 FSA produced 33 feasible solutions and a subset of those are shown here, including the statistically optimal solution denoted in bold with $R^2 = 0.1256308$ (column 3)

Variable 1	Variable 2	Times chosen by FSA	R^2
mb104327194	mb91638370	42	0.0957401
mb13136127	mb31255782	898	0.1245719
mb28636979	**mb87344525**	**107**	**0.1256308**
mb111935889	mb43233761	25	0.1065257
mb62443411	mb99541026	23	0.1088855
mb112250554	mb96331482	56	0.1123864

Columns 1 and 2 show the SNPs that were identified in each of the models

be found in the supplemental materials.) Out of the 2284 replications of FSA, the statistically optimal solution was identified in 107 of the replications, showing that the number of random starts used here was sufficient.

5 Discussion and Conclusion

Although FSA addresses limitations of existing modeling techniques, little is known about its theoretical properties. To address one aspect of this limitation, we have provided a way to determine the number of iterations of FSA needed to obtain the statistically optimal solution of an m-way interaction model with a certain probability in Theorem 1. For example, when considering a two-way interaction model, if you would like the probability of obtaining the statistically optimal solution to be at least 80%, then you would need to choose the number of random starts of FSA to be 40% of the number of possible explanatory variables in your data set. This lower bound on the probability of obtaining the statistically optimal solution can be easily implemented by data analysts running FSA. Further, simulation study and real data analysis demonstrated the validity and usefulness of this lower bound.

The work here provides a foundation for further study of theoretical properties of FSA. For instance, the simulation study results show that the derived lower bound is conservative. Thus, in future work, we aim to tighten the bound. However, in this case, the conservative lower bound does provide statistical guidance to FSA users. In addition, little is known about how conservative this bound is in the presence of smaller effect sizes, which will increase the impact of this work by providing more specific guidance to the data analyst. Knowing how to choose the number of times to run FSA will improve the computational usability of FSA by allowing the user to choose fewer random starts based on the desired likelihood of obtaining the statistically optimal solution while still being computationally feasible, and continue providing a valid alternative to exhaustive search methods.

References

Friedman, J., Hastie, T., Tibshirani, R.: glmnet: Lasso and elastic-net regularized generalized linear models. R package version 1(4) (2009)

Gemperline, P.J.: Computation of the range of feasible solutions in self-modeling curve resolution algorithms. Anal. Chem. **71**(23), 5398–5404 (1999)

Goudey, B., Abedini, M., Hopper, J.L., Inouye, M., Makalic, E., Schmidt, D.F., Wagner, J., Zhou, Z., Zobel, J., Reumann, M.: High performance computing enabling exhaustive analysis of higher order single nucleotide polymorphism interaction in genome wide association studies. Health Inf. Sci. Syst. **3**(1), 1 (2015)

Hawkins, D.M.: The feasible set algorithm for least median of squares regression. Comput. Stat. Data Anal. **16**(1), 81–101 (1993)

Hawkins, D.M.: A feasible solution algorithm for the minimum volume ellipsoid estimator in multivariate data. Comput. Stat. **8**, 95–95 (1993)

Hawkins, D.M.: The feasible solution algorithm for least trimmed squares regression. Comput. Stat. Data Anal. **17**(2), 185–196 (1994)

Hawkins, D.M.: The feasible solution algorithm for the minimum covariance determinant estimator in multivariate data. Comput. Stat. Data Anal. **17**(2), 197–210 (1994)

Hawkins, D.M., Olive, D.J.: Improved feasible solution algorithms for high breakdown estimation. Comput. Stat. Data Anal. **30**(1), 1–11 (1999)

Lambert, J., Gong, L., Elliot, C.F., Thompson, K., Stromberg, A.: rFSA: an R package for finding best subsets and interactions. R J. **10**(2), 295–308 (2018)

Lumley, T., Miller, A.: Leaps: regression subset selection. R package version 2 (2004)

Miller, A.J.: Selection of subsets of regression variables. J. R. Stat. Soc. Ser. A Gen. **147**(3), 389–425 (1984)

Moore, J.H., Williams, S.M.: Epistasis and its implications for personal genetics. Am. J. Hum. Genet. **85**(3), 309–320 (2009)

Zhang, W., Korstanje, R., Thaisz, Staedtler, F., Harttman, N., Xu, L., Feng, M., Yanas, L., Yang, H., Valdar, W., et al.: Genome-wide association mapping of quantitative traits in outbred mice. G3: Genes Genomes Genetics **2**(2), 167–174 (2012)

A Primer of Statistical Methods for Classification

Rajarshi Dey and Madhuri S. Mulekar

1 Introduction

Classification is a process of assigning a new subject or item to one of the G known groups or classes on the basis of how closely specific characteristics of this subject/item match with those of the groups. For example, on the basis of specific protein levels measured for a patient, an oncologist can determine with certain confidence whether or not that patient has a certain type of cancer; using a pixel-based satellite image, a geographer can classify land cover into different categories such as water, forested wetland, and upland forest; or using certain admissions criteria, a university can classify applicants as accepted or non-accepted into their program.

There are many different methods (or rules) used to achieve this goal of classification of subjects/items into different classes. Note that classification is not to be confused with clustering as classification involves assigning items to a known number of groups with specific characteristics, whereas clustering involves forming groups of items with similar characteristics when the existing number of groups is unknown. In a world of machine learning and computation, the process of classification is referred to as a supervised learning whereas the method of clustering is an example of unsupervised learning. For example, consider a chicken farm that packages chicken eggs. Before packaging, each egg has to be classified as medium, large, extra-large, or jumbo. This is a case of classification as the classes are well defined based on the egg size and machines can be taught to properly classify eggs based on their size. Sometimes clustering or other pattern recognition methods are

R. Dey · M. S. Mulekar (✉)

Department of Mathematics and Statistics, University of South Alabama, Mobile, AL, USA

e-mail: rajarshidey@southalabama.edu; mmulekar@southalabama.edu

© Springer Nature Switzerland AG 2019

N. Diawara (ed.), *Modern Statistical Methods for Spatial and Multivariate Data*,

STEAM-H: Science, Technology, Engineering, Agriculture, Mathematics & Health,

https://doi.org/10.1007/978-3-030-11431-2_6

used by scientists as a first step towards classification in which existence of number of groups is determined using multiple characteristics of its constituents. In this article, we discuss classification only in the sense of supervised learning, i.e., a procedure in which discriminating variables or functions are used to predict group membership.

Classification into one of the two groups (i.e., $G = 2$) is known as a binary classification and is relatively easier to implement. The earlier classification procedures developed were mostly binary classification methods. But with the development of computing and technology, more literature on different aspects of classification has become available (Hastie et al. 2001).

Broadly speaking there are two types of classifiers: hard and soft. As Wahba (2002) described, the soft classifiers also known as probabilistic classifiers typically provide conditional probability of membership for each of G ($G \geq 2$) groups for each new subject to be classified and then put the subject in the group with the largest probability of membership. In contrast, hard classifiers only provide a hard classification boundary like a fence around a property for each group based on the explanatory variables. A new subject with characteristics within the boundary of a group is assigned to that particular group.

The development of a classification procedure involves two major steps, classification and validation. In the first step, a classification rule or an algorithm also known as a classifier is developed and in the second step, performance of this classifier is evaluated. Dataset used in the first step to develop a classification rule is known as the *training dataset* and includes information about group membership (response or output) of each subject along with other random variables (explanatory variables or features). A similar dataset is used in the second step to validate the classifier developed in the first step and is known as the *validation dataset*.

Let us assume that all outcomes are independently observed for random response variable Y and a vector of random explanatory variables $\mathbf{X} = (\mathbf{X}_1, \mathbf{X}_2, \ldots, \mathbf{X}_p)$. Let $\{(\mathbf{x}_i, y_i) : i = 1, 2, \ldots n\}$ indicate a training dataset consisting of n independent measurements. Each measurement is a $(p + 1)$-tuple where $\mathbf{x}_i = (x_{1i}, x_{2i}, \ldots, x_{pi})'$ is a vector of outcomes for p explanatory variables and y_i ($i = 1, 2, \ldots, n$) is the outcome for response variable, which only shows membership of ith unit to a certain class.

Consider the famous iris dataset by R.A. Fisher (Anderson 1935) containing 150 data points with four explanatory variables and three classes of iris. The explanatory variables are sepal-length, sepal width, petal-length, and petal-width of three types of irises, namely *setosa, virginica,* and *versicolor.* The goal is to identify type of iris using sepal and petal measurements. This is a classification problem. Sepal-length and petal-length distributions in Fig. 1 show that although *setosa* tend to have lowest and *virginica* tend to have highest sepal-lengths the separation among three classes based on sepal-length is not clear due to considerable overlap and hence sepal-length by itself is not a good classifier for these three types of irises. On the other hand, petal-length is clearly able to distinguish *setosa* from the other two but not between *virginica* and *versicolor* possibly resulting in misclassification. Hence there is need for more than one predictor to reduce misclassification. Figure 2

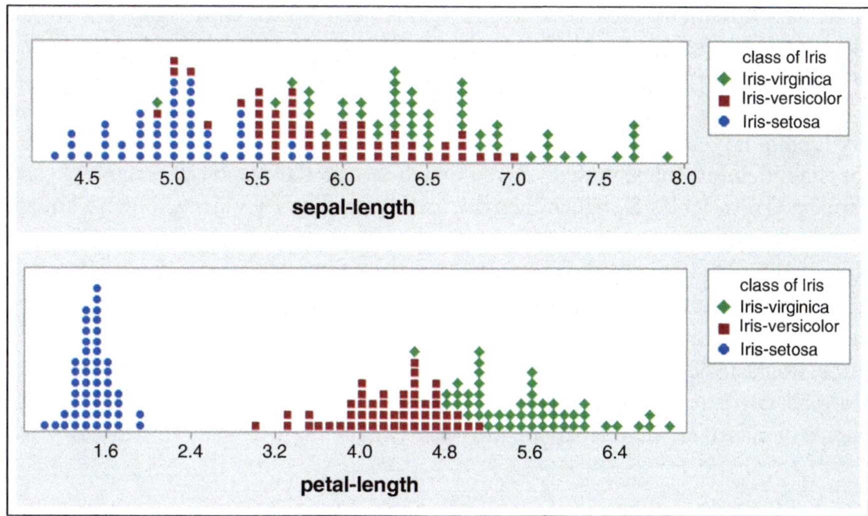

Fig. 1 Distributions of sepal- and petal-lengths of three varieties of iris

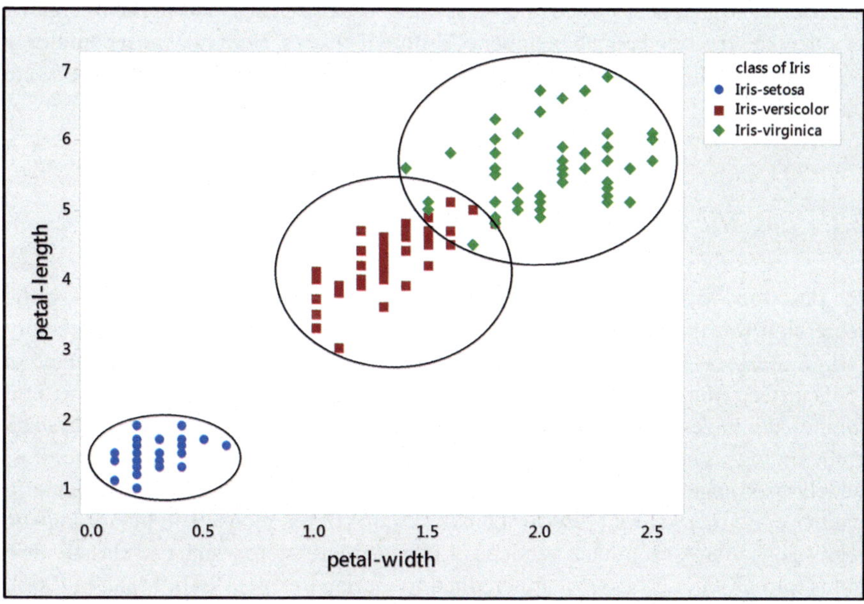

Fig. 2 A scatterplot of iris petal-length vs petal-width

shows a scatterplot of iris petal-length versus petal-width. Once again *setosa* are clearly separated from the other two varieties, but there is still some overlap between *virginica* and *versicolor* possibly leading to misclassification.

In this article, we aim to provide basic description of the most well-known and commonly used classification methods that are used to develop classifiers (or classification rules) based on relation between the response variable Y and explanatory variables $\mathbf{X}$, which then are used to assign new objects to these known groups based on observed $\mathbf{x}_0'$. Two soft classifiers (logistic regression and naïve Bayes estimator) and four hard classifiers (linear discriminant analysis, support vector machines, K nearest neighbor, and classification trees), respectively, are described in Sects. 2 and 3 along with their strengths and weaknesses. Some discussion assessing performance of these classifiers for five different datasets, three real and two simulated, is provided in Sect. 4. Some concluding remarks about choice of classifiers in practice are provided in Sect. 5.

2 Soft Classifiers

Intuitively, a soft classifier should appeal to anyone who likes to incorporate the uncertainty of outcome provided by classifiers because it also shows the likelihood of a new observation being a member of different classes. Here two most commonly used soft classifiers, namely logistic regression and naïve Bayes classifiers are discussed.

2.1 Logistic Regression

As described by Cramer (2003), the first use of logistic function in logistic regression was traced to modeling population growth rate in the nineteenth century Africa. Berkson (1944) suggested the use of logistic probability density function (pdf) instead of normal pdf in certain bioassay procedures. He also coined the term *logit* model to describe the resulting model. Later many researchers in statistics and epidemiology started working on what would eventually become one of the most widely used methods in classification, namely the logistic regression, particularly with $G = 2$ groups. Cox (1969) is considered one of the pioneers in binary logistic regression. More generalized versions of logistic regression, which can classify new items into G ($G \geq 2$) groups, are credited to Gurland et al. (1960), Mantel (1966), and Theil (1969).

For the sake of simplicity, let's start with the case of binary logistic regression with two classes being coded as 1 and 2 (i.e., $y_i = 1$ or 2). For $\pi_{i1} = P(y_i = 1 | \mathbf{x}_i)$, $i = 1, 2, \ldots, n$ and under the assumption that the response variable y_i has a

Bernoulli distribution with parameters π_{i1}, the logistic model is given by,

$$\pi_{i1} = E\left(y_i = 1|\mathbf{x}_i\right) = \frac{\exp\left(\beta_0 + \sum_{j=1}^{p}\beta_j x_{ji}\right)}{1 + \exp\left(\beta_0 + \sum_{j=1}^{p}\beta_j x_{ji}\right)} \tag{1}$$

where $\beta_0, \beta_1, \ldots, \beta_p$ are $(p+1)$ regression coefficients. Note that in a binary case, $\pi_{i2} = 1 - \pi_{i1}$.

Alternatively this model can be presented as,

$$logit\left(\pi_{i1}\right) = \log\left(\frac{\pi_{i1}}{1 - \pi_{i1}}\right) = \beta_0 + \sum_{j=1}^{p}\beta_j x_{ji} \tag{2}$$

The regression parameters $\beta_j, j = 0, 1, \ldots, p$ are estimated from the available training dataset. The maximum likelihood (ML) estimates $\hat{\beta}_j, j = 0, 1, \ldots, p$ are obtained by maximizing the likelihood function

$$L\left(\beta_0, \beta_1, \ldots, \beta_p\right) = \prod_{i=1}^{n} \pi_i^{y_i}(1 - \pi_i)^{n-y_i}.$$

Since there are no closed-form solutions available for maximizing this likelihood function, iterative algorithms are used to obtain the ML estimates of regression parameters. According to Agresti (2013), the most popular choices for iterative algorithms are either the Newton–Raphson algorithm (Tjalling 1995) or iteratively reweighted least square (IRWLS) algorithm (Burrus et al. 1994). However, sometimes due to the use of too many explanatory variables or highly correlated explanatory variables, these algorithms fail to converge resulting in failure to estimate parameters. Another counter-intuitive situation sometimes occurs when there is a complete separation between two classes using some linear combination of explanatory variables. More information on estimating parameters of logistic regression is available in Menard (2002).

A simple extension of logistic regression from binary to multiclass classification is known as multinomial logistic regression. The multinomial logistic model is given by,

$$\pi_{ig} = \frac{\exp\left(\beta_{0g} + \sum_{j=1}^{p}\beta_{jg} x_{ji}\right)}{1 + \sum_{g=1}^{G-1} \exp\left(\beta_{0g} + \sum_{j=1}^{p}\beta_{jg} x_{ji}\right)}, \quad g = 1, 2, \ldots, G-1 \text{ and } i = 1, 2, \ldots, n. \tag{3}$$

Extending notation used in the binary case to $G \geq 2$ groups, we can write $\pi_{ig} = P(y_i = g|\mathbf{x}_i)$, for $i = 1, 2, \ldots, n$ and $g = 1, 2, \ldots, G$. Although it does not matter which category is chosen as baseline, generally category G is used as a baseline and π_{iG} can be obtained using the fact that $\pi_{iG} = 1 - \sum_{g=1}^{G-1} \pi_{ig}$. From the point of estimation, there are $(p+1)(G-1)$ model parameters to be estimated. For estimating these parameters, the ML estimation or the maximum a posteriori (MAP) methods are commonly used (Murphy 2012). Estimation method MAP is similar to ML in the sense that it chooses that value of parameter which maximizes the value of a mathematical function, in this case the posterior distribution of the parameter itself. Most of the times, a closed-form solution is not available, hence different algorithms are used for estimation and IRWLS is a popular choice among practitioners.

If the $G \geq 2$ classes are ordered using an ordinal response variable, an alternative popular model often used in practice is the proportional-odds cumulative logit model. For example, consider a typical Likert scale question where the responders are asked to grade certain experience on a scale of 1 to 5 with 1 being the worst rating and 5 being the best. It might be of interest to determine if there exist some explanatory variables that can explain how the responders rate their experience. First developed by Snell (1964), this model is given by,

$$L_{ig} = \log \frac{\sum_{c=1}^{g} \pi_{ic}}{\sum_{c=g+1}^{G} \pi_{ic}} = \beta_{0g} + \sum_{j=1}^{p} \beta_j x_{ji}, \quad \text{for } g = 1, 2, \ldots, G \text{ and } i = 1, 2, \ldots, n.$$

$$(4)$$

Here, L_{ig} represents the log-odds of two cumulative probabilities. A manageable number of total $(G - 1 + p)$ parameters are to be estimated from this model. Typically, ML estimates of parameters of this model are obtained using iterative algorithms such as IRWLS and majorization-minimization (Lange 2016).

2.2 Naïve Bayes Classifier

Naïve Bayes (NB) is a family of soft classifiers that uses the Bayes theorem (Bayes 1763) along with a very strong assumption of independence among explanatory variables which is often unrealistic. However, this classifier works very well in the presence of dependencies among many categorical explanatory variables (Rish 2001) and is quite fast to execute even with large datasets.

NB classifier differs from the logistic regression classifier in terms of how the probability π_{ig} is modeled. When using a logistic regression classifier, $\pi_{ig} = P(y_i = g \mid \mathbf{x}_i)$ is modeled directly from data. On the other hand, when using a Naïve Bayes classifier, first the estimates for $P(\mathbf{x}_i \mid y_i = g)$ are obtained from data and then assuming independence among explanatory variables, π_{ig} is modeled using Bayes theorem as,

$$\pi_{ig} \propto P(y_i = g) \prod_{j=1}^{p} P(x_{ji} \mid y_i = g), \quad g = 1, 2, \ldots, G \text{ and } i = 1, 2, \ldots, n. \quad (5)$$

The estimate for $P(y_i = g)$ can be obtained from the training set as the proportion of training set observations that belong to class g ($g = 1, 2, \ldots, G$). The estimates for $P(x_{ji}|y_i = g)$ are typically obtained via ML estimation technique. A new observation is assigned to a group for which probability π_{ig} is maximum among all G groups.

Estimating parameters from the likelihood function depends on how the likelihood, $P(x_{ji}|y_i = g)$, $i = 1, 2, \ldots, n$, $j = 1, 2, \ldots, p$, and $g = 1, 2, \ldots, G$ is modeled parametrically. If X_j is a continuous random variable, then the popular choice of distribution is normal (Gaussian) such that $(X_j| Y = g) \sim N(\mu_g, \Sigma_g)$. If X_j is a categorical random variable with m categories, then the most commonly used distribution is multinomial, i.e., $(X_j| Y = g) \sim Multinomial(1, \phi_{1g}, \ldots, \phi_{mg})$ for one trial where ϕ_{lg}, $l = 1, 2, \ldots m$ is the probability associated with the lth category such that $\sum_{l=1}^{m}\phi_l = 1$.

The NB classifier is remarkably effective considering the assumptions needed to obtain the probabilities are almost always wrong (Hand and Yu 2001). This method is a building block to what is commonly known as a Bayesian spam filter (Nigam et al. 2000) used by the email providers. A semi-parametric version of NB classifier performs much better when the explanatory variables are obviously non-normal (Soria et al. 2011).

3 Hard Classifiers

Hard classifiers typically do not provide a probability of group association. In other words, there is no uncertainty associated with classification because classifier provides a hard boundary between groups and exactly for this reason some researchers like to use them. Four commonly used classifiers discussed here are linear discriminant analysis, K nearest neighbor, support vector machines, and classification trees.

3.1 Linear Discriminant Analysis

Linear discriminant analysis (LDA) is a hard classification method. Statistical literature indicates that LDA is one of the first methods developed for classification and its basic idea originated from none other than Fisher (1936). The basic idea behind LDA is to determine that linear combination of explanatory variables which will magnify the difference between two classes making it easier to achieve correct classification. The generalization of this idea for classification into $G(G > 2)$ classes is credited to Rao (1948). The NB classifier is similar to LDA in nature (Hand and Yu 2001), although in LDA the aim is obtain a classifier while in NB there is more emphasis on identifying a class with the maximum posterior probability.

Fisher (1936) proposed a classification rule for two groups which involved determining a vector $\mathbf{r}$ that maximizes function $\delta(\mathbf{r}) = (\mathbf{r}'\Sigma\mathbf{r})^{-1}(\mathbf{r}'(\mu_1 - \mu_2))^2$

under the assumption that $(\mathbf{X}|\,Y = g)\sim N(\boldsymbol{\mu}_g, \boldsymbol{\Sigma})$ for $g = 1, 2$. This is equivalent to finding a hyperplane that provides a solution to equation

$$\log\left[\frac{P\,(Y = 1|\mathbf{X})}{P\,(Y = 2|\mathbf{X})}\right] = 0. \tag{6}$$

Using Bayes' rule, we can write,

$$P\,(Y = g|\mathbf{X}) = P\,(Y = g)\,\frac{P\,(\mathbf{X}|Y = g)}{P\,(\mathbf{X})} \quad \text{for } g = 1, 2$$

where $p_g = P(Y = g)$ for $g = 1, 2$ is the overall class probability and can be estimated from the training data. Under the assumption that the explanatory variables are multivariate normal, the hyperplane can be found by solving the following equation for $\mathbf{r}$,

$$\log\left[\frac{p_1}{p_2}\right] + \mathbf{r}'\boldsymbol{\Sigma}^{-1}\,(\boldsymbol{\mu}_1 - \boldsymbol{\mu}_2) - \frac{1}{2}\left(\boldsymbol{\mu}_1'\boldsymbol{\Sigma}^{-1}\boldsymbol{\mu}_1 - \boldsymbol{\mu}_2'\boldsymbol{\Sigma}^{-1}\boldsymbol{\mu}_2\right) = 0. \tag{7}$$

Solution to (7) leads to a *linear* classifier (or a linear boundary between two groups) because Eq. (6) is a linear function of explanatory variables. The first step in LDA is to estimate the mean vectors ($\boldsymbol{\mu}_1$ and $\boldsymbol{\mu}_2$) and variance–covariance matrix ($\boldsymbol{\Sigma}$) from the training dataset. For any new observation, $\mathbf{x}_0$, one can estimate $\Delta_{\mathbf{x}_0}$ from (8) as,

$$\hat{\Delta}_{\mathbf{x}_0} = \log\left[\frac{\hat{p}_1}{\hat{p}_2}\right] + \mathbf{x}_0'\hat{\boldsymbol{\Sigma}}^{-1}\,(\hat{\boldsymbol{\mu}}_1 - \hat{\boldsymbol{\mu}}_2) - \frac{1}{2}\left(\hat{\boldsymbol{\mu}}_1'\hat{\boldsymbol{\Sigma}}^{-1}\hat{\boldsymbol{\mu}}_1 - \hat{\boldsymbol{\mu}}_2'\hat{\boldsymbol{\Sigma}}^{-1}\hat{\boldsymbol{\mu}}_2\right). \tag{8}$$

Using this $\hat{\Delta}_{\mathbf{x}_0}$ value a new observation $\mathbf{x}_0$ is assigned to one of the two groups as follows:

$$\text{Assign } \mathbf{x}_0 \text{ to } \begin{cases} \text{Group 1} & \text{if } \hat{\Delta}_{\mathbf{x}_0} > 0 \\ \text{Group 2} & \text{if } \hat{\Delta}_{\mathbf{x}_0} < 0. \end{cases}$$

In cases where the assumption of homoscedasticity of variance–covarinace matrix is not justified and a more general underlying assumption is that $(\mathbf{X}|\,Y = g)\sim N(\boldsymbol{\mu}_g, \boldsymbol{\Sigma}_g)$ for $g = 1, 2$, a *quadratic* classifier (i.e., a quadratic function) is used to describe a boundary between two classes. This procedure is known as quadratic discriminant analysis (QDA) (Hastie et al. 2001). In QDA, the hyperplane can be obtained by solving (9) for $\mathbf{r}$.

$$\begin{aligned} \log\left[\frac{p_1}{p_2}\right] - \tfrac{1}{2}\log\frac{|\boldsymbol{\Sigma}_1|}{|\boldsymbol{\Sigma}_2|} + \mathbf{r}'\left(\boldsymbol{\Sigma}_1^{-1}\boldsymbol{\mu}_1 - \boldsymbol{\Sigma}_2^{-1}\boldsymbol{\mu}_2\right) \\ - \tfrac{1}{2}\mathbf{r}'\left(\boldsymbol{\Sigma}_1^{-1} - \boldsymbol{\Sigma}_2^{-1}\right)\mathbf{r} - \tfrac{1}{2}\left(\boldsymbol{\mu}_1'\boldsymbol{\Sigma}_1^{-1}\boldsymbol{\mu}_1 - \boldsymbol{\mu}_2'\boldsymbol{\Sigma}_1^{-1}\boldsymbol{\mu}_2\right) = 0 \end{aligned} \tag{9}$$

As can be seen from (7) and (9), a QDA requires more parameters to be estimated from the training dataset, precisely $(2 + 2p + 2p^2)$ parameters for QDA compared to $(2 + 2p)$ for LDA. That can lead to a serious issue if the training dataset is small. To overcome this issue, Srivastava et al. (2007) proposed an effective Bayesian solution.

A simpler method under the assumption of homoscedasticity of variance–covariance matrices is to use Mahalanobis distance (Mahalanobis 1936) for classification. For any new observation, $\mathbf{x}_0$, a linear discriminant function LDF_g is computed for each group (see (10)) under the assumption that $\boldsymbol{\mu}_g$ and $\boldsymbol{\Sigma}$, respectively, are the unknown mean vector and variance–covariance matrix of $\mathbf{X}$.

$$LDF_g(\mathbf{x}_0) = \mathbf{x}_0' \hat{\boldsymbol{\Sigma}}_0^{-1} \hat{\boldsymbol{\mu}}_g - \frac{1}{2} \hat{\boldsymbol{\mu}}_g' \hat{\boldsymbol{\Sigma}}_0^{-1} \hat{\boldsymbol{\mu}}_g + \hat{p}_g \tag{10}$$

where $\hat{\boldsymbol{\Sigma}}_0$ is the pooled estimate of the common variance–covariance matrix $\boldsymbol{\Sigma}$ and $\hat{p}_g$ is the estimate of probability $p_g = P(Y = g)$ for $g = 1, 2, \ldots, G$ obtained from the training data. Then the new observation $\mathbf{x}_0$ is assigned to the group with the highest discriminant function value, i.e., the group corresponding to $\max\{LDF_g(\mathbf{x}_0), g = 1, 2, \ldots, G\}$.

Although LDA is quite effective in many situations (Hand 2006), in some situations the joint pdf of explanatory variables differs considerably from the multivariate normal distribution. In such cases semi-parametric LDA technique derived by Mai and Zou (2015) under the assumption of sparse variance–covariance matrix is more effective.

3.2 K Nearest Neighbor

Assumed to have originated in long past, the history of $K(1 < K < n)$ nearest neighbor (KNN) classification is not really that well known. In modern times, Sebestyen (1962) described this method as *proximity algorithm* and Nilsson (1965) called it the *minimum distance classifier*. Cover and Hart (1967) were the first to name this algorithm as the *nearest neighbor* and that name became popular.

Although mostly used as a hard classifier, KNN can be used as a soft classifier too. The idea behind KNN is quite simple and no parametric assumption is required. Given a training dataset of size $n(n > K)$, this classification algorithm starts when a new observation, $\mathbf{x}_0 = (x_{01}, \ldots, x_{0p})'$, is recorded with known values for all explanatory variables but unknown class. The first step is to calculate the K nearest neighbors in terms of the explanatory variables. Using some well-defined distance measure, distance $d_i = d(\mathbf{x}_0, \mathbf{x}_i)$, $i = 1, 2, \ldots, n$ between this new observation and each observation from the training dataset is calculated and these distances are ordered as $d_{(1)} \leq d_{(2)} \leq \ldots \leq d_{(n)}$. Considering the lowest K distances, $\{d_{(1)}, d_{(2)}, \ldots, d_{(K)}\}$, the class membership of these closest K neighbors in the training dataset is determined. Then the new observation is placed in the class that has the largest number of these K neighbors. For example, suppose k_g of the nearest K neighbors belong to group $g(g = 1, 2, \ldots, G)$ such that $\sum_{g=1}^{G} k_g = K$, then the new observation is placed in the group c if $k_c = \max \{k_g, g = 1, 2, \ldots, G\}$. Note that there is a possibility that no such unique maximum exists for a given new observation and a chosen K, thus resulting in ties. Although not exactly a group inclusion probability, these nearest neighbors can be used to provide a group membership indicator of the new observation using relative fractions (k_g/K), $g = 1, 2, \ldots, G$.

Now the question is: how to choose value of K, the number of nearest neighbors to be used? Given a large dataset one can always use cross-validation and choose the K value corresponding to the lowest misclassification rate in the validation dataset. Note that choice of a too small value for K indicates that the space generated by the explanatory variables is divided into many small subspaces and the class membership of a new observation depends on which subspace the new observation belongs to. In that case outliers in the original dataset can create problems in predicting the class membership of a new observation that is close to the outlier resulting in a higher variance in prediction. However, choice of a large value for K basically leads to division of the training data space into G smooth subspaces which in turn creates the problem of misclassification of any outlier of these subspaces and subsequently higher bias in prediction. As a rule of thumb, $K = \sqrt{n}$ is considered to be a sensible choice for number of classes in practice. If the number of groups in the data is 2 (i.e., $G = 2$), then K should be an odd number to avoid the possibility of ties in group membership indicators.

The most popular choice for a distance measure is the Euclidean distance which for a new observation, $\mathbf{x}_0 = (x_{01}, \ldots, x_{0p})'$, is calculated as,

$$d_i = \sqrt{\sum_{j=1}^{p} \left(x_{0i} - x_{ji}\right)^2}, \quad i = 1, 2, \ldots, n. \tag{11}$$

Some other distance measures used commonly in practice are Hamming distance (Hamming 1950) and Chebyshev distance (Grabusts 2011). Chomboon et al. (2015) looked at eleven different distance measures and found that the Euclidean, Chebyshev, and Mahalanobis distance measures perform well. For a synopsis of different distance measures, please refer to Mulekar and Brown (2014). Different explanatory variables tend to have different range of possible values and some distance measures such as the Euclidean distance tend to be affected by the range of measurements. Hence in practice, datasets are typically normalized before classification to reduce the influence of explanatory variables with larger range of measurements. When using a dataset with a large number of explanatory variables, to reduce the computation time, a dimension reduction technique such as principal component analysis (PCA) is used. To overcome the problem of choosing a value for K, Samworth (2012) suggested the use of weighted nearest neighbor algorithm in which instead of choosing K nearest neighbors (i.e., essentially assigning a weight of $1/K$ to K nearest neighbors and 0 to the remaining observations in the training dataset while assigning a class to the new observation), all observations in the training dataset are assigned a weight using some optimal weighting scheme. When dealing with a big dataset, an approximation to the method of nearest neighbors proposed by Har-Peled et al. (2012) is useful.

3.3 *Support Vector Machine*

Support vector machine (SVM) is a class of hard classifiers. For a binary classification with p explanatory variables, an SVM classifier constructs a $(p - 1)$-dimensional hyperplane in the pth dimension to maximize the margin. Here margin refers to the distance between the observation closest to the boundary of a group and the remaining groups. Points on or closest to the boundary of decision surface are called support vectors and they are used in learning models associated with the classification algorithm. The idea behind SVM is to find that hyperplane which provides the maximum margin from support vectors among infinitely many possible hyperplanes that can separate two groups provided the two groups are completely separable. For a binary classification, one hyperplane known as the maximum margin hyperplane is constructed. For $G > 2$ groups, more than one such maximum margin hyperplanes need to be created to separate groups and a combination of these hyperplanes is used for the classification of a new observation.

Consider the case of binary classification, and assume that there actually exists a linear hyperplane of the form

$$W\left(\mathbf{X}\right) = w_0 + \sum_{j=1}^{p} w_j X_j \tag{12}$$

that can perfectly differentiate between two classes. Then a method described by Vapnik and Lerner (1963) can be used to find a maximum margin hyperplane. Maximum margin hyperplane is a hyperplane for which $W(\mathbf{X}) = 0$. In SVM, only support vectors obtained using the training data are used to estimate the coefficients of explanatory variables in (12). Since the decision surface differentiates the classes completely, the linear function in (12) should be positive for one group and negative for another. Without any loss of generality, assume that for support vector(s) in group 1, $\hat{W}(X) = -1$ and for those in group 2, $\hat{W}(X) = 1$. In order to maximize the margin, it is sufficient to minimize $\sum_{j=1}^{p} w_j^2$ subject to $v_i \hat{W}\left(\mathbf{x}_i\right) \geq 1, i = 1, 2, \ldots, n$ where $v_i = -1$ if $y_i = 1$ and $v_i = 1$ if $y_i = 2$. Thus this hyperplane can be obtained by minimizing the Lagrangian formulation,

$$L = -\sum_{i=1}^{n} a_i \left(v_i W\left(\mathbf{x}_i\right) - 1\right)$$

where $a_i(\ i = 1, 2,\ \ldots, n)$ are Lagrange multipliers. Once this hyperplane is estimated, $\hat{W}(\mathbf{x}_0)$ is computed for any new observation $\mathbf{x}_0$ and the new observation is assigned to the group 1 if the $\hat{W}(\mathbf{x}_0) < 0$ and to group 2 if $\hat{W}(\mathbf{x}_0) > 0$.

In many practical situations, a perfectly differentiating hyperplane does not exist. For such situations, Cortes and Vapnik (1995) proposed a modification to the maximum margin hyperplane to differentiate between two groups. They proposed estimating the hyperplane with the help of a hinge loss function, $h(\mathbf{x}) = \max\left(0, 1 - v\hat{W}(\mathbf{x})\right)$. Note that unlike a linearly separable case where $v_i\hat{W}(\mathbf{x}_0) \geq 1\ \forall i$; for linearly non-separable cases, the possibility of $v_i\hat{W}(\mathbf{x}_0) < 0$ exists for a few support vectors. So, the hinge loss function is 0 for such support vectors and this is used to penalize such support vectors while estimating the decision boundary. Thus, instead of minimizing $\sum_{j=1}^{p} w_j^2$ subject to $v_i\hat{W}(\mathbf{x}_i) \geq 1, i = 1, 2, \ldots, n$, function

$$\theta \sum_{j=1}^{p} w_j^2 + \frac{1}{n} \sum_{i=1}^{n} \max\left(0, 1 - v_i\hat{W}(x_i)\right) \tag{13}$$

is minimized where θ is a penalty parameter. This is known as the *soft* version of SVM, although this is not a soft classifier.

Since a linear classifier does not always exist, researchers extended the idea of SVM to find a non-linear classifier. Using ideas first promoted by Aizerman et al. (1964), Boser et al. (1992) proposed the use of kernelization to obtain a non-linear classifier by improving a classifier obtained via SVM. The trick is to use a transformation $\mathbf{Z}$ of $\mathbf{X}$ such that the new transformed explanatory variables $\mathbf{Z}(\mathbf{X})$ provide a better classifier than the one provided by $\mathbf{X}$. Then, proceed to obtain an SVM based on the new transformed explanatory variables, $\mathbf{Z}$. A carefully chosen transformation $\mathbf{Z}$ can possibly result in a linear classifier. Note that $\mathbf{Z}$ is not observed or calculated from data but it is replaced by the kernel function, κ, such that $\kappa(\mathbf{x}_i, \mathbf{x}_l) = \mathbf{z}_i'\mathbf{z}_l$ for $i \neq l = 1, 2, \ldots, n$. There are many kernel functions in use, but the most used Gaussian kernel (Schölkopf et al. 1997) is given by,

$$\kappa(x_i, x_l) = \exp\left(-\xi \sum_{j=1}^{p} (x_{ij} - x_{lj})^2\right).$$

To develop a multiclass SVM classifier (i.e., for $G > 2$), there are few options available. In a method known as *one-against-all*, G SVM classifiers are obtained for each class separately and a new observation is assigned to the class chosen by maximum number of these classifiers (Bottou et al. 1994). In another method known as *one-against-one* (Kressel 1998), $G(G - 1)/2$ SVM classifiers each separating a pair of classes are obtained and a new observation is assigned to the class that is predicted by the most classifiers (Kressel 1998). Hsu and Lin (2002) who compared their performances concluded that *one-against-one* performs better than *one-against-all* in most of the situations that they studied. There are many modifications of SVM proposed by researchers from different fields that work better in certain specific situations. Typically, SVM works really well if there exists a good separation between classes or when the number of explanatory variables is large compared to the sample size of the training dataset. SVM is not computationally effective when using a very large training dataset.

3.4 Classification Trees

Classification trees (CT) are methods used to partition the space of explanatory variables into disjoint subsets and assign a class to each subset by minimizing some measure of misclassification also known as impurity. It is a visually pleasing method and can be easily as well as effectively described to those from the non-scientific communities. CT can handle large datasets as well as missing data, and it can easily ignore bad explanatory variables. However, sometimes depending on the dataset CT can produce a really bad partition of the space of explanatory variables leading to high misclassification rates.

CT produces a flowchart or tree structure starting with a root node (one explanatory variable) and then, proceeds with splits (internal nodes) until no split is deemed necessary (leaf nodes). Each leaf node is assigned to a class. There are many algorithms on how to select a root node, how to split a node, how many splits of each node are needed, and when to stop splitting a node to make it a leaf node. An example of classification tree is shown in Fig. 3. It shows classification of a random sample of $n = 78$ from iris data by R.A. Fisher (Anderson 1935) using JMP 12 into one of the three classes using two explanatory variables petal-width and petal-length.

The root node is typically chosen with an explanatory variable that provides the lowest rate of misclassification. This is easily achieved when the number of explanatory variables is small. For example, let there be two classes ($G = 2$) and one explanatory variable ($p = 1$). Consider the rule for using two complementary subgroups A_g created by a split with the explanatory variable such that, $y_i = g$ if $x_{1i} \in A_g$, $g = 1, 2$. For each split a Gini impurity measure is computed as,

$$I(CT) = \sum_{g=0}^{1} \left(1 - \sum_{j=0}^{1} \widehat{q_g(j)} \right)$$

where

$$\widehat{q_g(j)} = \frac{\sum_{i=1}^{n} I\left(y_i = j; x_{1i} \in A_g\right)}{\sum_{i=1}^{n} I\left(x_{1i} \in A_g\right)}$$

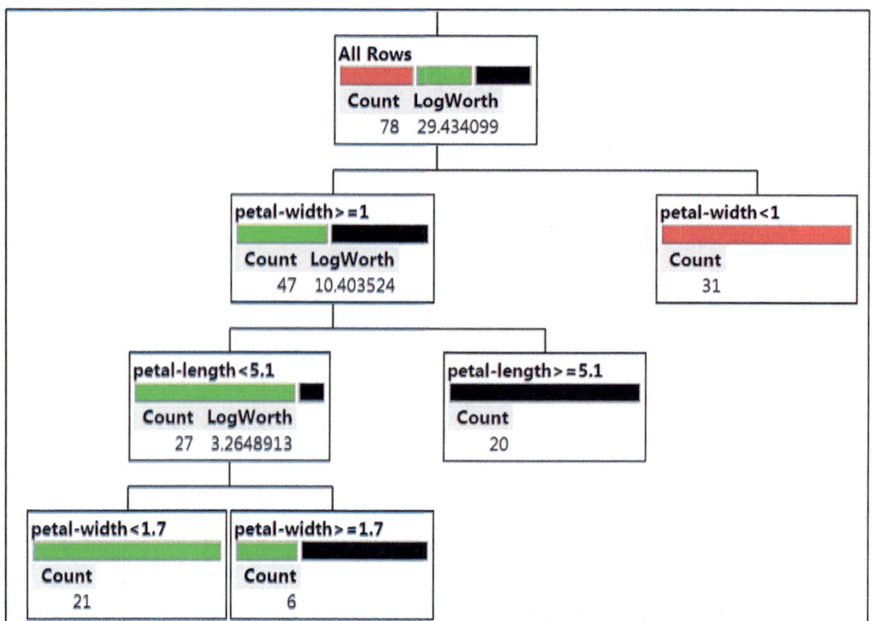

Fig. 3 Classification tree using a random sample of iris data ($n = 78$)

is the misclassification rate for group g and the splits are chosen such that the Gini impurity measure is minimized (Witten et al. 2011). As shown in Fig. 3, LogWorth for each model defined as $-\log_{10}(p\text{-value})$ is also another measure used to decide where to split. The p-value can be based on a chi-square test for a split that maximizes LogWorth value.

The first classification tree algorithm was proposed by Messenger and Mandell (1972). However, Breiman et al. (1984) have provided what became the most popular classification tree algorithm, namely classification and regression trees (CART). Several improved versions of it were proposed later and are still used in practice. Many modifications of the CART method have been proposed for various reasons, but mainly because CART produces biased and high-variance trees, i.e., changing the training set can drastically change the tree diagram. A few Bayesian versions of this algorithm are also available in the literature (Chipman et al. 1998; Denison et al. 1998). Loh (2009) provides a classification algorithm, called GUIDE which is computationally faster and incorporates nearest neighbor algorithm to improve the CT.

To reduce the variance in CT, two new methods were proposed, namely the bagging method (Breiman 1996) and the random forests method (Breiman 2001). Note that these methods do not produce a tree diagram but they focus on obtaining many classification trees from the training data so that a new observation that needs to be classified is put into the class suggested by majority of these trees. Bagging is simply achieved by obtaining bootstrap samples with replacement from the training data. Random forests are similar to bagging but in each tree only a randomly chosen subset of typically $\sqrt{p}$ number of explanatory variables is considered when determining the nodes.

Freund and Schapire (1997) introduced the concept of boosting which aims to reduce bias in a CT. This is achieved by refitting the data into trees with higher weights for misclassified data points. In the initial calculation of the first CT, all data points are given equal weight. However, the weights are updated after each iteration and the impurity measure is updated by assigning higher weights to the misclassified observations. Then the final classifier is selected via weighted average of the trees.

4 Assessment and Comparison of the Performance of Classifiers

The performance of a classifier is typically judged by cross-validating the classification rule with a separate dataset of size s, called the validation data. Sometimes cross-validation is also used to estimate unknown parameters such as the number of neighbors to be considered in KNN method. In the absence of a separate validation data, the idea of Jackknife sampling (Quenouille 1949, 1956; Tukey 1958) is used to obtain a K-fold cross-validation. In this special case of cross-validation, the training dataset is divided into K smaller datasets of equal size, and $(K - 1)$ of them are used as the training data and the remaining K^{th} one as the validation dataset. This process is repeated K times until each of them is used once as the validation data.

The simplest performance measure of a classifier is the misclassification rate $R(0 \leq R \leq 1)$, which is the proportion of validation sample that is misclassified. Hence a small value of $R(R \to 0)$ is an indication of more accurate classification. Although a very simple metric, this is an effective measure of performance. It works well as long as the cost of misclassification for and sample sizes from all classes are relatively similar. If sample sizes differ considerably, then the use of an uncertainty coefficient is recommended (Mills 2011). Uncertainty coefficient U is calculated as $U = (H - H_c)/H \ (0 \leq U \leq 1)$ where

$$H = -\sum_{g=1}^{G} P\left(Y = g\right) \log\left(P\left(Y = g\right)\right)$$

and

$$H_c = -\sum_{g=1}^{G} \sum_{l=1}^{G} P\left(Y = g, \hat{Y} = l\right) \log\left(P\left(Y = g | \hat{Y} = l\right)\right)$$

can be estimated from the training data. Larger values of U indicate a better classifier. If the accuracy of classification for only one class is very important, then one can calculate the sensitivity (also known in medicine as the true positive rate or in machine learning as the recall rate) for that class. Sensitivity also takes values between 0 and 1 but a good classifier is expected to have a higher sensitivity. The sensitivity for class g can be calculated as,

$$sen_g = \frac{\sum_{i=1}^{n} I\,(y_i = g, \widehat{y}_i = g)}{\sum_{i=1}^{n} I\,(y_i = g, \widehat{y}_i = g) + \sum_{i=1}^{n} I\,(y_i = g, \widehat{y}_i \neq g)}$$

where I is the indicator variable taking values 1 or 0 depending on whether the condition is satisfied or not.

Besides misclassification rate, sensitivity, and uncertainty coefficient, there are many other performance measures available to judge classification methods, a detailed discussion of which is provided by Hand (2012).

Some articles dedicated to comparison of different classification methods are available. The earlier research was mostly focused on comparing logistic regression against LDA (Hosmer et al. 1983 and McLachlan and Byth 1979). Their outcomes indicate that LDA is a better performer if the explanatory variables are normally distributed but the advantage diminishes as sample size becomes larger. Meshbane and Morris (1996) recommended that QDA should be used instead of LDA if the distributions of explanatory variables are skewed. After comparing outcomes using classification tree and KNN, Liu and White (1995) concluded that KNN performs better than classification tree unless the number of explanatory variables is large. A study by Bhattacharya et al. (2011) compared SVM to logistic regression for detecting credit card fraud and found no advantage in using more complicated method like SVM over simpler logistic regression. Finch and Schneider (2006) compared performances of logistic regression, discriminant analysis, and classification trees based on simulated data while Kiang (2003) compared performances of logistic regression, LDA, KNN, and classification tree based on a separate set of simulated data. Asparoukhov and Krzanowski (2001) compared performances of all but one (namely SVM) classifiers mentioned in this paper using five real-life datasets for binary classification. They also discussed the effect of choosing different sized training set along with changing the number of explanatory variables. Steel et al. (2000) argue that simply comparing the methods is not completely meaningful unless the model selection process (i.e., the choice of explanatory variables in the final prediction model) is included. The only conclusion that can be drawn from all these studies is that there is no winner among all methods that work in every situation very effectively. The performance of a classifier depends very much on the dataset for which a classification is needed.

In this section, we compare the performance of six classifiers discussed earlier using two simulated datasets and three real-life datasets with respect to the misclassification rates and uncertainty measure U. All computations were done using available R packages *rpart*, *e1071*, *class*, *naivebayes*, and *MASS*. Of the two simulated datasets, one is visually separable albeit not linearly while the other is

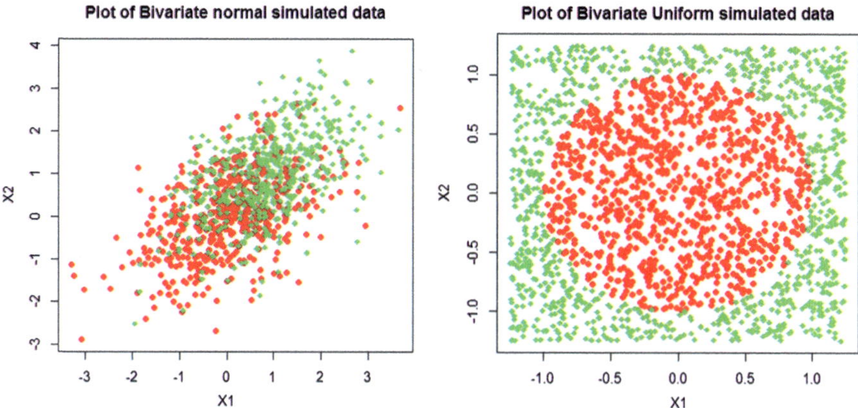

Fig. 4 Visualization of the simulated datasets

Table 1 Comparison of misclassification rates for different classifiers for five datasets

	Iris	Skin	Glass	Bivariate normal	Bivariate uniform
Logistic	0.09804	0.0401	NA	0.288	0.495
LDA	0.03922	0.0505	0.2364	0.300	0.495
NB	0.03922	0.0245	0.4545	0.304	0.077
KNN	0.03922	0.0048	0.1818	0.002	0.000
SVM	0.05882	0.0051	0.3273	0.312	0.008
CT	0.07843	0.0232	0.2545	0.320	0.096

Table 2 Comparison of uncertainty measure for different classifiers for five datasets

	Iris	Skin	Glass	Bivariate normal	Bivariate uniform
Logistic	0.71463	0.72119	NA	0.12654	0.00116
LDA	0.86424	0.71310	0.52570	0.11903	0.00116
NB	0.88589	0.80570	0.43788	0.11408	0.66483
KNN	0.88589	0.94258	0.62021	0.96238	1.00000
SVM	0.79531	0.93916	0.25449	0.06982	0.93602
CT	0.78020	0.80141	0.52453	0.10053	0.54546

not separable using any reasonable curve and provides a challenge in terms of classification (see Fig. 4). Both datasets have two explanatory variables as presented in scatterplots in Fig. 4 where each class is represented by separate point type and color. For NB classifier, Gaussian prior was used. For KNN classifier, the next larger odd integer to $\sqrt{n}$ was used as the value of K, except in one real data example (skin data), where this value was too large due to large dataset. The observed misclassification rates for six methods and five examples are listed in Table 1 and the uncertainty coefficients are listed in Table 2.

Example 1 (Iris) Consider the famous iris dataset by R.A. Fisher (Anderson 1935) described in the Introduction. Fifty observations are available for each type of iris. Of the 150 measurements available, a training dataset of 99 observations was created with 33 observations each from three groups. The misclassification rate was estimated based on the remaining 51 observations that constituted a validation sample. Misclassification rates lower than 0.10 (Table 1) and uncertainty measures over 0.70 (Table 2) show that all the methods did a commendable job of correct classification for this data. However, NB, KNN, and LDA are slightly better than other classifiers.

Example 2 (Skin) Refer to the skin segmentation dataset from the UCI machine learning repository (Bhatt et al. 2009). This dataset contains 245,057 observations randomly sampled from photos of faces of people of different age group, gender, and color. Of those, 50,589 observations are for samples of skin while the rest are for samples of non-skin parts of the face. The three explanatory variables in this example are RGB triplet, i.e., red, green, and blue colors used in displaying images. RGB values are typically given as an integer value in the range of 0–255, and combined together they determine the color of the image which in this case is part sampled. Distribution of RGB pixels for skin data is presented in a 3-dimensional plot in Fig. 5. Without rotating the plot around three axes it is difficult to tell if there is clear distinction or some overlap between two groups, skin and non-skin. A training sample of size 150,000 was used, out of which 30,000 were skin samples.

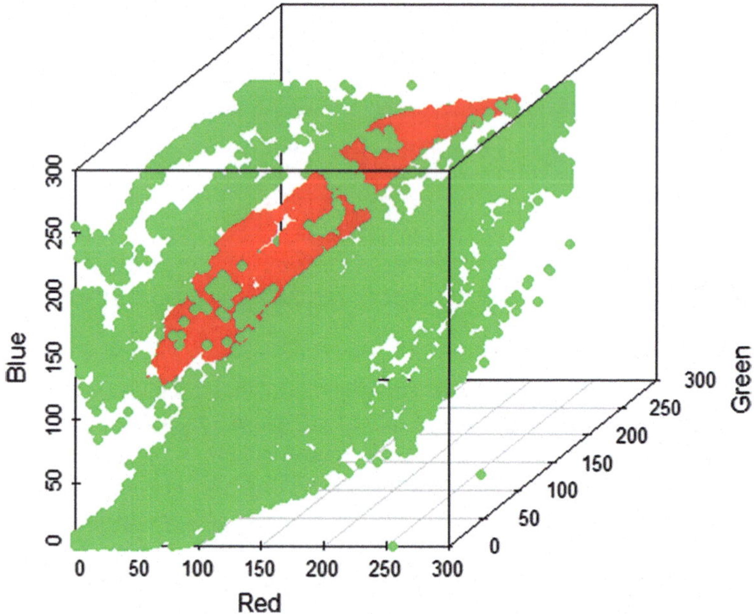

Fig. 5 Visualization of RGB pixel distribution for skin data

Tables 1 and 2 show that KNN and SVM perform best for this data, followed by NB and CT. To save computation time, $K = 19$ was used for KNN algorithm.

Example 3 (Glass) Consider the glass dataset from UCI machine learning repository (Lichman 2013). The original dataset describes six types of glass samples along with the refractive index and weight percent of oxides formed with sodium, magnesium, aluminum, silicon, potassium, calcium, barium, and iron in the sample. Since some of the classes have small sample sizes, only three types of glass were used for classification purpose in this example. They are float-processed building window glasses (70 measurements), nonfloat-processed building window glasses (76 measurements), and non-window headlamp glasses (29 measurements). Fifty samples each from two classes of building window glasses and 20 samples from headlamp were used as the training data. Simulations to obtain parameters for a multinomial logistic regression failed due to non-convergence of iterations. Outcomes in Tables 1 and 2 indicate that KNN performs the best followed by CT and LDA.

Example 4 (Bivariate Uniform) Consider two independent univariate uniform distributions, namely $X_i \sim Uniform(-1.25, 1.25)$ for $i = 1, 2$. A sample of 3000 observations was generated with seed 1234. With the unit circle providing the class boundary, the i-th observation is assigned to group 1 if $x_{1i}^2 + x_{2i}^2 \leq 1$ and to group 2 otherwise. The first 2000 observations generated were used as a training data and the remaining 1000 as a validation sample. In this training dataset, 1012 observations were from group 1 and the remaining 998 from group 2. In the validation dataset, 520 observations were from group 1 and the remaining 480 from group 2. Note that in this situation a linear classifier is not supposed to perform well because of non-normal distributions and that is reflected in the misclassification rates listed in Table 1 and uncertainty measures listed in Table 2. Logistic regression and LDA seems to be only as good as a coin toss in this situation whereas KNN and SVM perform admirably well.

Example 5 (Bivariate Normal) Now consider the bivariate normal populations. A sample of 1500 observations from two homoscedastic bivariate normal distributions that differ only in mean vector was generated using the *mvtnorm* package in R with seed 5678, resulting in a total sample of size 3000. The difference in the mean vectors and the variance–covariance matrix used in the simulation were, respectively,

$$\mu_1 - \mu_2 = \begin{bmatrix} 1.0 \\ 1.0 \end{bmatrix} \quad \text{and} \quad \Sigma = \begin{bmatrix} 1.0 & 0.5 \\ 0.5 & 1.0 \end{bmatrix}.$$

The first 1000 rows were used as a training sample for both groups (resulting in a total sample size of 2000) and the remaining 500 as the validation sample (resulting total sample size 1000). Tables 1 and 2 show KNN as a clear winner while the other classifiers are almost equally bad. Although we expected LDA to perform better, to our surprise the results say otherwise.

5   Concluding Remarks

In this paper, the basic ideas that dominate the world of statistical classification were described. Detailed discussions of them are scattered in different textbooks, but none discusses them all together. For example, logistic regression is discussed in detail by Kleinbaum and Klein (2010), LDA by McLachlan (2004), SVM by Steinwart and Christmann (2008), classification trees by Breiman et al. (1984), and different classification methods by Izenman (2008) and James et al. (2013).

For data with highly correlated explanatory variables or a large number of explanatory variables, the use of some dimension reduction technique such as principal component analysis, low variance filter, and high correlation filter before classification is recommended (Farcomeni and Greco 2015). In cases where $p > n$, dimension reduction becomes necessary. Alternatively, although random forests method is not a dimension reduction technique for explanatory variables, in cases where $\sqrt{p} < n$ this method can be effectively used without reducing dimension of explanatory variables.

A very basic question on this topic should be about the preference for any particular classification method. Alternatively, should there be preference for a certain classification method over the others. It depends on the circumstances. There is no single method that stands out as the best. Typically for complex problems in which the misclassification rate is higher among all classifiers, the use of soft classifiers is recommended. However, hard classifiers remain popular as their outcomes are easier to interpret in practice. Also hard classifiers like SVM and KNN generally provide good outcomes as seen from the situations discussed here. In this age of computation, the most recent research emphasis is on effective ways of implementing bagging and random forests (James et al. 2013) which can be computationally more effective than other classifiers like KNN. Liu et al. (2011) describe a suave large-margin unified machine that combines margin-based hard and soft classifiers, and that hard classifiers tend to perform better than soft classifiers when the classes are either easily separable or when the training sample size is relatively small compared to number of explanatory variables.

Research over the years has led to the development of many classifiers. As a result, the toolbox from which a classifier can be chosen provides an extensive list of options which to some extent depends on software used by and the computing power available for the researcher. Also comparative performance of different classifiers is changing with changing technology and results of past studies might lead to different conclusions with the current technology. Thus, one can entertain the idea of using all possible classifiers and assign a new observation to a class assigned by most of the classifiers.

References

Agresti, A.: Categorical Data Analysis. Wiley, Hoboken (2013)

Aizerman, M.A., Braverman, E.M., Rozonoer, L.I.: Theoretical foundations of the potential function method in pattern recognition learning. Autom. Remote. Control. **25**, 821–837 (1964)

Anderson, E.: The irises of the Gaspe Peninsula. Bull. Am. Iris Soc. **59**, 2–5 (1935)

Asparoukhov, O.K., Krzanowski, W.J.: A comparison of discriminant procedures for binary variables. Comput. Stat. Data Anal. **38**, 139–160 (2001)

Bayes, T.: An essay towards solving a problem in the doctrine of chances. Philos. Trans. **53**, 370–418 (1763)

Berkson, J.: Applications of the logistic function to bioassay. J. Am. Stat. Assoc. **9**, 357–365 (1944)

Bhatt, R.B., Sharma, G., Dhall, A., Chaudhury, S.: Efficient Skin Region Segmentation Using Low Complexity Fuzzy Decision Tree Model. IEEE-Indicon, Ahmedabad (2009)

Bhattacharya, S., Sanjeev, J., Tharakunnel, K., Westland, J.C.: Data mining for credit card fraud: a comparative study. Decis. Support. Syst. **50**, 602–613 (2011)

Boser, B.E., Guyon, I.M., Vapnik, V.N.: A training algorithm for optimal margin classifiers. In: Proceedings of the Fifth Annual Workshop on Computational Learning Theory – COLT '92. p. 144 (1992)

Bottou, L., Cortes, C., Denker, J.S., Drucker, L., Guyon, I., Jackel, L., LeCun, Y., Muller, U.A., Sackinger, E., Simard, P., Vapnik, V.N.: Comparison of classifier methods: a case study in handwriting digit recognition. Int. Conf. Pattern Recognit. **2**, 77–87 (1994)

Breiman, L.: Bagging predictors. Mach. Learn. **24**(2), 123–140 (1996)

Breiman, L.: Random forests. Mach. Learn. **45**, 5–32 (2001)

Breiman, L., Friedman, J.H., Olshen, R.A., Stone, C.J.: Classification and Regression Trees. Wadsworth, Belmont (1984)

Burrus, C.S., Barreto, J.A., Selesnick, I.W.: Iterative reweighted least squares design of FIR filters. IEEE Trans. Signal Process. **42**(11), 2922–2936 (1994)

Chipman, H.A., George, E.I., McCulloch, R.E.: Bayesian CART model search. J. Am. Stat. Assoc. **93**, 935–948 (1998)

Chomboon, K., Pasapichi, C., Pongsakorn, T., Kerdprasop, K., Kerdprasop, N.: An empirical study of distance mreics for K-nearest neighbor algorithm. 3rd International Conference on Industrial Application Engineering, 280–285 (2015)

Cortes, C., Vapnik, V.: Support-vector networks. Mach. Learn. **20**(3), 273–297 (1995)

Cover, T., Hart, P.: Nearest neighbor pattern classification. IEEE Trans. Inf. Theory. **13**(1), 21–27 (1967)

Cox, D.R.: Analysis of Binary Data. Chapman and Hall, London (1969)

Cramer, J.S.: The origins and development of the logit model. In: Cramer, J.S. (ed.) Logit Models from Economics and Other Fields, pp. 149–158. Cambridge University Press, Cambridge (2003)

Denison, D.G.T., Mallick, B.K., Smith, A.F.M.: A Bayesian CART algorithm. Biometrika. **85**, 363–377 (1998)

Farcomeni, A., Greco, L.: Robust Methods for Data Reduction. CRC Press, Boca Raton (2015)

Finch, W.H., Schneider, M.K.: Misclassification rates for four methods of group classification. Educ. Psychol. Meas. **66**(2), 240–257 (2006)

Fisher, R.A.: The use of multiple measurements in taxonomic problems. Ann. Eugenics. **7**(2), 179–188 (1936)

Freund, Y., Schapire, R.E.: A decision-theoretic generalization of on-line learning and an application to boosting. J. Comput. Syst. Sci. **5**(1), 119–139 (1997)

Grabusts, P.: The choice of metrics for clustering algorithms. Proceedinhs of the 8th International Scientific and Practical Conference, **11**, 70–76 (2011)

Gurland, J., Lee, I., Dahm, P.A.: Polychotomous quantal response in biological assay. Biometrics. **16**, 382–398 (1960)

Hamming, R.W.: Error detecting and error correcting codes. Bell Syst. Tech. J. **29**(2), 147–160 (1950)

Hand, D.J.: Classifier technology and the illusion of progress. Stat. Sci. **21**, 1–14 (2006)

Hand, D.J.: Assessing the performance of classification methods. Int. Stat. Rev. **80**, 400–414 (2012)

Hand, D.J., Yu, K.: Idiot's Bayes - not so stupid after all? Int. Stat. Rev. **69**(3), 385–399 (2001)

Har-Peled, S., Indyk, P., Motwani, R.: Approximate nearest neighbor: towards removing the curse of dimensionality. Theory Comput. **8**, 321–350 (2012)

Hastie, T., Tibshirani, R., Friedman, J.: The Elements of Statistical Learning: Data Mining, Inference and Prediction. Springer, New York (2001)

Hosmer, T., Hosmer, D.W., Fisher, L.L.: A comparison of the maximum likelihood and discriminant function estimators of the coefficients of the logistic regression model for mixed continuous and discrete variables. Commun. Stat. **12**, 577–593 (1983)

Hsu, C.W., Lin, C.J.: A comparison of methods for multi-class support vector machines. IEEE Trans. Neural Netw. **13**(2), 415–425 (2002)

Izenman, A.J.: Modern Multivariate Statistical Techniques: Regression, Classification and Manifold Learning. Springer, New York (2008)

James, G., Witten, D., Hastie, T., Tibshirani, R.: An Introduction to Statistical Learning: With Applications in R. Springer, New York (2013)

Kiang, M.: A comparative assessment of classification methods. Decis. Support. Syst. **35**, 441–454 (2003)

Kleinbaum, D.G., Klein, M.: Logistic Regression: A Self-learning Text, 3rd edn. Springer, New York (2010)

Kressel, U.H.G.: Pairwise classification and support vector machines. In: Advances in Kernel Methods: Support Vector Learning, pp. 255–268. MIT Press, Cambridge (1998)

Lange, K.: MM Optimization Algorithms. SIAM, Philadelphia (2016)

Lichman, M.: UCI Machine Learning Repository [http://archive.ics.uci.edu/ml]. University of California, Irvine 2013

Liu, W.Z., White, A.P.: A comparison of nearest neighbor and tree-based methods of nonparametric discriminant analysis. J. Stat. Comput. Simul. **53**, 41–50 (1995)

Liu, Y., Zhang, H.H., Wu, Y.: Hard or soft classification? Large-margin unified machines. J. Am. Stat. Assoc. **106**(493), 166–177 (2011)

Loh, W.Y.: Improving the precision of classification trees. Ann. Appl. Stat. **3**, 1710–1737 (2009)

Mahalanobis, P.C.: On the generalized distance in statistics. Proceedings of the National Institute of Science in India, **2**(1), 49–55, (1936)

Mai, Q., Zou, H.: Semiparametric sparse discriminant analysis in ultra-high dimensions. J. Multivar. Anal. **135**, 175–188 (2015)

Mantel, N.: Models for complex contingency tables and polychotomous response curves. Biometrics. **22**, 83–110 (1966)

McLachlan, G.J.: Discriminant Analysis and Statistical Pattern Recognition. Wiley-Interscience, New York (2004)

McLachlan, G.J., Byth, K.: Expected error rates for logistic regression versus normal discriminant analysis. Biom. J. **21**, 47–56 (1979)

Menard, S.: Applied Logistic Regression Analysis, 2nd edn. Sage Publications, Thousand Oaks (2002)

Meshbane, A., Morris, J.D.: A method for selecting between linear and quadratic classification models in discriminant analysis. J. Exp. Educ. **63**(3), 263–273 (1996)

Messenger, R., Mandell, L.: A modal search technique for predictive nominal scale multivariate analysis. J. Am. Stat. Assoc. **67**, 768–772 (1972)

Mills, P.: Efficient statistical classification of satellite measurements. Int. J. Remote Sens. **32**, 6109–6132 (2011)

Mulekar, M.S., Brown, C.S.: Distance and similarity measures. In: Alhaji, R., Rekne, J. (eds.) Encyclopedia of Social Network and Mining (ESNAM), pp. 385–400. Springer, New York (2014)

Murphy, K.P.: Machine Learning: A Probabilistic Perspective. MIT Press, Cambridge (2012)

Nigam, K., McCallum, A., Thrun, S., Mitchell, T.: Text classification from labeled and unlabeled documents using EM. Mach. Learn. **39**(2-3), 103–134 (2000)

Nilsson, N.: Learning Machines: Foundations of Trainable Pattern-Classifying Systems. McGraw-Hill, New York (1965)

Quenouille, M.H.: Problems in plane sampling. Ann. Math. Stat. **20**(3), 355–375 (1949)

Quenouille, M.H.: Notes on bias in estimation. Biometrika. **43**(3-4), 353–360 (1956)

Rao, R.C.: The utilization of multiple measurements in problems of biological classification. J. R. Stat. Soc. Ser. B. **10**(2), 159–203 (1948)

Rish, I.: An empirical study of the naive Bayes classifier. In: IJCAI Workshop on Empirical Methods in AI, Sicily, Italy (2001)

Samworth, R.J.: Optimal weighted nearest neighbour classifiers. Ann. Stat. **40**(5), 2733–2763 (2012)

Schölkopf, B., Sung, K., Burges, C., Girosi, F., Niyogi, P., Poggio, T., Vapnik, V.: Comparing support vector machines with Gaussian kernels to radial basis function classifiers. IEEE Trans. Signal Process. **45**, 2758–2765 (1997)

Sebestyen, G.S.: Decision-making Process in Pattern Recognition. McMillan, New York (1962)

Snell, E.J.: A scaling procedure for ordered categorical data. Biometrics. **20**, 592–607 (1964)

Soria, D., Garibaldi, J.M., Ambrogi, F., Biganzoli, E.M., Ellis, I.O.: A non-parametric version of the naive Bayes classifier. Knowl.-Based Syst. **24**(6), 775–784 (2011)

Srivastava, S., Gupta, M.R., Frigyik, B.A.: Bayesian quadratic discriminant analysis. J. Mach. Learn. Res. **8**, 1287–1314 (2007)

Steel, S.J., Louw, N., Leroux, N.J.: A comparison of the post selection error rate behavior of the normal and quadratic linear discriminant rules. J. Stat. Comput. Simul. **65**, 157–172 (2000)

Steinwart, I., Christmann, A.: Support Vector Machines. Springer, New York (2008)

Theil, H.: A multinomial extension of the linear logit model. Int. Econ. Rev. **10**(3), 251–259 (1969)

Tjalling, J.Y.: Historical development of the Newton-Raphson method. SIAM Rev. **37**(4), 531–551 (1995)

Tukey, J.W.: Bias and confidence in not quite large samples. Ann. Math. Stat. **29**(2), 614–623 (1958)

Vapnik, V., Lerner, A.: Pattern recognition using generalized portrait method. Autom. Remote. Control. **24**, 774–780 (1963)

Wahba, G.: Soft and hard classification by reproducing Kernel Hilbert space methods. Proc. Natl. Acad. Sci. **99**, 16524–16530 (2002)

Witten, I., Frank, E., Hall, M.: Data Mining. Morgan Kaufmann, Burlington (2011)

A Doubly-Inflated Poisson Distribution and Regression Model

Manasi Sheth-Chandra, N. Rao Chaganty, and Roy T. Sabo

1 Introduction

The Poisson distribution is the standard choice for modeling probabilities of count data and can be modeled against covariates using the generalized linear models framework (McCulloch and Searle 2001). A common extension of these models is to account for over-dispersion, or situations where the count variability exceeds the count mean, see Cameron and Trivedi (1998) for a summary of approaches to verify and analyze over-dispersed count data. Another extension of the basic Poisson model is to account for excess zeros. Both Cohen (1963) and Johnson and Kotz (1969) describe the zero-inflated Poisson (ZIP) distribution, and Lambert (1992) extended this distribution to account for covariates that could simultaneously model the counts as well as the probability of particular zeros being in excess. More details on Poisson regression models and zero-inflated models are provided by Hall (2000), Bae et al. (2005), Coxe et al. (2009), and Hall and Shen (2010).

One example of count data is a patient's length of stay (LOS) before hospital discharge, which is often used as a simple count measure of the cost of care (Marazzi et al. 1998). The use of LOS as a measure of cost assumes that a patient with a larger LOS is of more cost to a hospital than a patient with a shorter LOS.

M. Sheth-Chandra
Center for Global Health, Old Dominion University, Norfolk, VA, USA
e-mail: msheth@odu.edu

N. R. Chaganty (✉)
Department of Mathematics and Statistics, Old Dominion University, Norfolk, VA, USA
e-mail: rchagant@odu.edu

R. T. Sabo
Department of Biostatistics, Virginia Commonwealth University, Richmond, VA, USA
e-mail: rsabo@vcu.edu

© Springer Nature Switzerland AG 2019

N. Diawara (ed.), *Modern Statistical Methods for Spatial and Multivariate Data*,
STEAM-H: Science, Technology, Engineering, Agriculture, Mathematics & Health,
https://doi.org/10.1007/978-3-030-11431-2_7

Table 1 Observed length of stay (LOS)

LOS	Frequency	LOS	Frequency	LOS	Frequency
0	55	5	20	10	3
1	35	6	13	11	1
2	35	7	8	12	4
3	75	8	4	13	0
4	40	9	5	14	1

This measure is a natural candidate for a ZIP model, as many patients are served in outpatient scenarios for which LOS is recorded as zero ($LOS = 0$). However, other frequencies can be inflated as well. Consider the data presented in Table 1, which shows a sample LOS data. While the frequency of zero-valued LOS is certainly high, we notice that $LOS = 3$ is also inflated relative to all other values. This may be due to patients staying three nights at the hospital to receive inpatient treatment followed by recovery time. As another example, consider the dental epidemiology data presented in Bohning et al. (1997, 1999). Here, the *decayed, missing, filled teeth* (DMFT)-index is used to measure the dental status of 1013 school children of age 7, which is a count of the number of decayed, missing, or filled teeth. This study focused on the eight deciduous molars, and the DFMT was measured at baseline and at 1 year. The goal of the study was to compare the absolute change in DMFT ($\delta(DMFT) = |DMFT_1 - DMFT_2|$) between four types of decay prevention, their combination, and a control, in the presence of other covariates. In this case zero was inflated as most of the children exhibited no change in dental status, and as such the zero count—corresponding to no improvement and/or consistent dental care—was inflated. However, $\delta(DFMT)$ scores of one were also inflated. The count of one is for children that showed improvement in only one cavity.

In both the length of stay and dental epidemiology examples, there were inflated instances of count $k > 0$ *in addition* to an inflation of zeros. As ZIP models can only account for excess zeros, here we introduce a doubly-inflated Poisson (DIP) distribution. This distribution can probabilistically model excessive zeros as well as excess counts at some positive integer value k. This model can also be adapted to the generalized linear models framework to account for covariates in a regression-like setting in a similar manner as the ZIP model. The added advantage of the DIP is that it can model covariates against the preponderance of both excess zeros and k values in addition to the count outcome. Lin and Tsai (2013) studied excessive zero and k responses in the context of health survey data. However, they did not consider method of moments estimation and comparisons with maximum likelihood using asymptotic relative efficiency criteria. They also did not derive analytically the elements for Fisher information as we did in this paper.

This chapter is organized as follows. The DIP distribution is presented in Sect. 2. In addition to the distributional form and parameterization of the DIP model, likelihood and moment-based estimation processes are outlined and their asymptotic efficiencies are compared. The hospital LOS data are used to exemplify this distribution. The DIP model is extended to the generalized linear model framework in Sect. 3, where models are developed for subject-specific (ungrouped) and frequency (grouped) data. These models are then used to analyze the dental epidemiology data. A brief summary concludes the manuscript in Sect. 4, including an outline of additional extensions of the DIP model. Finally, the Appendix contains some analytical first and second- order partial derivatives of the log-likelihood functions.

2 Doubly-Inflated Poisson Distribution

2.1 *Parameterization*

The doubly-inflated Poisson model can be constructed as a mixture of binomial $(2, p)$ and Poisson (λ) distributions where $0 < p < 1$ and $\lambda > 0$, which can be interpreted as a mixture of a Poisson process and two degenerate distributions with masses concentrated at zero and at a positive integer value k. The probability mass function of a $DIP(p, \lambda)$ random variable Y can be written as:

$$f(y; p, \lambda) = \begin{cases} p^2 + q^2 f_1(0; \lambda), & \text{for } y = 0; \\ 2pq + q^2 f_1(k; \lambda), & \text{for } y = k; \\ q^2 f_1(y; \lambda), & \text{for } y = 1, 2, \dots \neq k. \end{cases} \tag{1}$$

where $q = 1 - p$ and $f_1(y; \lambda) = \lambda^y \exp(-\lambda)/(y!)$ is the Poisson mass function with mean λ. This simple choice of modeling two probabilities with one parameter and its compliment guarantees identifiability. Properties of the distribution (1) can be found in Sheth-Chandra (2011). Based on (1), the probabilities of inflated instances of $y = 0$ and $y = k$ outside the standard Poisson are p^2 and $2pq$, respectively. The probabilities for all counts governed by the standard Poisson distribution are then scaled downward by q^2. Note that as $p \to 0$ this model reduces to the ordinary Poisson distribution. A doubly-inflated model could be created with separate proportions p_1 and p_2 to account for inflations at $y = 0$ and $y = k$ respectively, but the additional parameter adds considerable complexity to both the model parameterization and estimation and as such this model will not be considered here.

2.2 *Methods of Estimation*

Maximum Likelihood

Assuming our data consists of independent count responses $y_i,\ i = 1, \dots, n$, generally governed by a Poisson(λ) distribution but with inflated counts at 0 and k.

The log-likelihood function of the $DIP(p, \lambda)$ model given in (1) can be written as

$$\ell(p, \lambda | y) = n_0 \log[f(0; p, \lambda)] + n_k \log[f(k; p, \lambda)] + \sum_{i:\, y_i \neq 0, k} \log[f(y_i; p, \lambda)],$$

$$(2)$$

where n_0 represents the number of zero counts and n_k represents the number of k counts. Differentiating log-likelihood function (2) with respect to the p and λ (assuming k is known) and solving the resulting score functions for those parameters yield the maximum likelihood (ML) estimates $(\widehat{p}, \widehat{\lambda})$. The Newton-Raphson algorithm can be used to find numerical solutions as the score equations are not in the closed form. The asymptotic variances $\sigma^2(\widehat{p})$ and $\sigma^2(\widehat{\lambda})$ of the maximum likelihood estimates $\widehat{p}$ and $\widehat{\lambda}$, respectively, are obtained by taking the diagonal elements of the inverse of the Fisher information matrix. The analytical formulas for the elements of the Fisher information matrix are given in Appendix "Information Matrix".

Method of Moments

To obtain moment estimates for p and λ, we note that the first two population moments are $\mu_1 = E(Y) = 2pqk + q^2\lambda$ and $\mu_2 = E(Y^2) = 2pqk^2 + q^2\left(\lambda + \lambda^2\right)$. We equate these to their corresponding sample moments $\overline{y}_1 = \frac{1}{n}\sum_{i=1}^{n} y_i$ and $\overline{y}_2 = \frac{1}{n}\sum_{i=1}^{n} y_i^2$, which yields the following two equations:

$$\begin{cases} \overline{y}_1 &= 2pqk + q^2\lambda \\ \overline{y}_2 &= 2pqk^2 + q^2\left(\lambda + \lambda^2\right). \end{cases} \quad (3)$$

Solving Eq. (3) numerically for p and λ using Newton-Raphson yields moment estimators $(\widetilde{p}, \widetilde{\lambda})$ for p and λ, respectively. If we let

$$D = \begin{pmatrix} \dfrac{\partial \mu_1}{\partial p} & \dfrac{\partial \mu_1}{\partial \lambda} \\ \dfrac{\partial \mu_2}{\partial p} & \dfrac{\partial \mu_2}{\partial \lambda} \end{pmatrix},$$

a matrix of first-order partial derivatives of the population moments, and letting Σ be the covariance matrix of Y and Y^2 given by

$$\Sigma = \begin{pmatrix} Var(Y) & Cov(Y, Y^2) \\ Cov(Y^2, Y) & Var(Y^2) \end{pmatrix}, \quad (4)$$

then the asymptotic covariance matrix for the moment estimators $(\widetilde{p}, \widetilde{\lambda})$ is given by $A = (D)^{-1} \Sigma (D^T)^{-1}$ (see Theorem A.1 in Chaganty and Shi (2004)). The diagonal elements of covariance matrix A are the asymptotic variances $\sigma^2(\widetilde{p})$ and $\sigma^2(\widetilde{\lambda})$ of the moment estimators $\widetilde{p}$ and $\widetilde{\lambda}$, respectively.

2.3 Asymptotic Relative Efficiencies

To compare the performance of the ML and moment estimators we calculate asymptotic relative efficiencies (ARE), taking the form of ratios of the asymptotic variances $e(\widetilde{p}, \widehat{p}) = \sigma^2(\widehat{p})/\sigma^2(\widetilde{p})$ and $e(\widetilde{\lambda}, \widehat{\lambda}) = \sigma^2(\widehat{\lambda})/\sigma^2(\widetilde{\lambda})$, respectively. Ratios less than 1 imply the ML estimators are more efficient than the moment estimators, ratios greater than 1 imply the moment estimators are more efficient than the ML estimators, and ratios close to 1 imply the two estimators are equally efficient.

Table 2 presents efficiencies for p $(e(\widetilde{p}, \widehat{p}))$ and λ $(e(\widetilde{\lambda}, \widehat{\lambda}))$ for a $DIP(p, \lambda)$ model with inflated 0's and 3's. The efficiencies are calculated for various values of $p \in (0, 1)$ and $\lambda \in [3, 9]$. Inspection of Table 2 shows an inverse relationship between the efficiencies for p and λ. For small values of p and λ—implying low rates of inflation and a low mean for the Poisson counts—the ML estimators are much more efficient than the moment estimators at estimating p than for λ. Conversely, the ML and moment estimators are nearly as efficient in estimating p for large values of p and λ, but the ML estimators are much more efficient in estimating λ than the moment estimators for those same scenarios. In general, since the efficiencies for estimating λ are nowhere greater than 0.8, the ML estimators can be said to outperform the moment estimators overall.

Table 3 presents AREs $e(\widetilde{p}, \widehat{p})$ and $e(\widetilde{\lambda}, \widehat{\lambda})$ for the DIP model with inflated 0's and 6's. A similar pattern to that shown in Table 2 is also shown here, with the efficiencies for p and λ having an inverse relationship for small and large values of

Table 2 Asymptotic relative efficiencies for DIP (p, λ) model for inflated 0's and 3's

p	$e(\widetilde{p}, \widehat{p})$				$e(\widetilde{\lambda}, \widehat{\lambda})$			
	$\lambda = 3$	$\lambda = 5$	$\lambda = 7$	$\lambda = 9$	$\lambda = 3$	$\lambda = 5$	$\lambda = 7$	$\lambda = 9$
0.1	0.0766	0.0859	0.4343	0.6629	0.7709	0.2219	0.5807	0.6988
0.2	0.2926	0.2592	0.6384	0.8174	0.2930	0.3895	0.6172	0.6973
0.3	0.0819	0.4404	0.7524	0.8854	0.2784	0.4869	0.6215	0.6712
0.4	0.2594	0.5853	0.8205	0.9223	0.4875	0.5419	0.6089	0.6302
0.5	0.4475	0.6953	0.8659	0.9454	0.5946	0.5746	0.5892	0.5829
0.6	0.6106	0.7808	0.9007	0.9614	0.6555	0.5951	0.5678	0.5352
0.7	0.7427	0.8499	0.9297	0.9735	0.6938	0.6086	0.5469	0.4907
0.8	0.8478	0.9075	0.9551	0.9835	0.7197	0.6180	0.5276	0.4495
0.9	0.9319	0.9569	0.9784	0.9921	0.7384	0.6248	0.5101	0.4132

Table 3 Asymptotic relative efficiencies for DIP (p, λ) model for inflated 0's and 6's

	$e(\widetilde{p}, \widehat{p})$				$e(\widetilde{\lambda}, \widehat{\lambda})$			
p	$\lambda = 3$	$\lambda = 5$	$\lambda = 7$	$\lambda = 9$	$\lambda = 3$	$\lambda = 5$	$\lambda = 7$	$\lambda = 9$
0.1	0.1877	0.0152	0.0023	0.1468	0.3659	0.6813	0.0178	0.3197
0.2	0.3803	0.0274	0.1170	0.3544	0.5086	0.6890	0.3013	0.4399
0.3	0.5413	0.1692	0.2978	0.5081	0.4549	0.6916	0.4411	0.4854
0.4	0.6653	0.3464	0.4619	0.6211	0.3350	0.6928	0.5071	0.5055
0.5	0.7617	0.5090	0.5969	0.7105	0.2449	0.6934	0.5442	0.5158
0.6	0.8363	0.6461	0.7074	0.7851	0.1866	0.6939	0.5676	0.5216
0.7	0.8939	0.7595	0.7991	0.8491	0.1486	0.6942	0.5837	0.5251
0.8	0.9382	0.8537	0.8764	0.9053	0.1229	0.6944	0.5953	0.5273
0.9	0.9727	0.9328	0.9426	0.9552	0.1047	0.6945	0.6041	0.5288

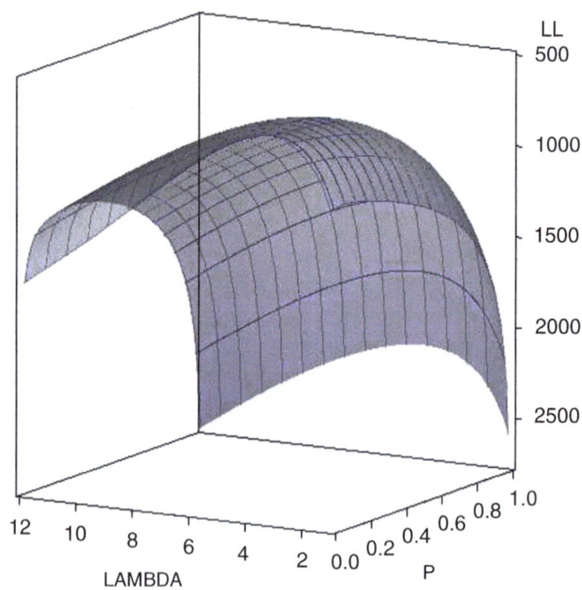

Fig. 1 Log-likelihood of LOS data using DIP model with parameters (p, λ)

the target parameters. However, in this case we see that the efficiencies for λ here are never larger than 0.7, implying that the moment estimators are not nearly as efficient as the ML estimators.

2.4 Hospital Length of Stay Example

Returning to the LOS data from Table 1, recall that there were inflated counts of patients who are received in outpatient settings ($n_0 = 55$), as well as inflated counts of patients receiving inpatient care for 3 days ($n_3 = 75$). This phenomenon of double inflation makes the $DIP(p, \lambda)$ a natural choice to model the inflation and count parameters. The log-likelihood for this model is presented in Fig. 1, with the peak matching the numerical ML estimates of $\widehat{p} = 0.161$ and $\widehat{\lambda} = 3.604$. The standard errors for the two estimates are 0.023 and 0.126, respectively, and the maximum log-likelihood is -666.0, at the ML estimates.

3 Doubly-Inflated Poisson Regression Modeling

In this section we allow parameters in the $DIP(p, \lambda)$ to be modeled by covariates in the generalized linear model framework McCulloch and Searle (2001). We focus on two general situations: one where the ith subject has a unique vector of l covariates $x_i = (x_{i1}, \ldots, x_{il})^T$, and another where the covariates occur in $s < n$ distinct combinations, such that several subjects are linked to the same covariate combination. In both cases, the covariates are separately linked to parameters p and λ via vectors of coefficients γ and β, respectively. Assuming canonical link functions for a binomial proportion and Poisson count, the structural component of the covariates and coefficients is modeled using logit and log link functions, respectively, such that for the ith subject we have

$$\text{logit}(p_i) = x_i^p \gamma = \gamma_0 + \gamma_1 x_{i1} + \ldots + \gamma_{m^p} x_{im^p}$$

$$(5)$$

$$\log(\lambda_i) = x_i^\lambda \beta = \beta_0 + \beta_1 x_{i1} + \ldots + \beta_{m^\lambda} x_{im^\lambda}$$

where x_i^p is the ith row of the set of covariates used to estimate p_i, and x_i^λ is the ith row of the set of covariates used to estimate λ_i. Here m^p and m^λ denote the number of covariates used for modeling p and λ, respectively, and these sets of covariates need not be identical for the two models in (5). Note $\exp(\gamma_l)$ is the multiplicative effect on the binomial probability from a 1-unit increase in x_{il} at the fixed levels of the other covariates. If x_{il} is a categorical indicator variable, then $\exp(\gamma_l)$ is the conditional odds ratio of the inflated counts. Similarly, a one-unit increase in x_{il} has a multiplicative impact of $\exp(\beta_l)$ on λ_i, and is equal to the relative risk if x_{il} is a categorical indicator.

3.1 Subject-Specific Data

When the count data $y_1, \ldots, y_n$ are accompanied by subject-specific sets of covariates (as shown in Table 4), the log-likelihood function (2) of the $DIP(p, \lambda)$

Table 4 General layout of raw count data

Subject	Response	Covariates		
1	y_1	x_{11}	$\ldots$	x_{1l}
2	y_2	x_{21}	$\ldots$	x_{2l}
$\vdots$	$\vdots$	$\vdots$	$\ddots$	$\vdots$
n	y_n	x_{n1}	$\ldots$	x_{nl}

regression model can be rewritten as

$$\ell(p_i, \lambda_i \mid y) = \sum_{\{i:y_i=0\}} \log\left(p_i^2 + q_i^2 \exp(-\lambda_i)\right) + \sum_{\{i:y_i=k\}} \log\left(2p_i q_i + q_i^2 \frac{\exp(-\lambda_i)\lambda_i^k}{k!}\right)$$

$$+ \sum_{\{i:y_i \neq 0,k\}} \log\left(q_i^2 \frac{\exp(-\lambda_i)\lambda_i^{y_i}}{y_i!}\right), \tag{6}$$

which can be reexpressed in terms of the regression parameters γ and β as

$$\ell(\gamma, \beta) = -2 \sum_{i=1}^{n} \log\left(1 + \exp(x_i^p \gamma)\right) + \sum_{\{i:y_i=0\}} \log\left(\exp(2x_i^p \gamma) + \exp(-\exp(x_i^\lambda \beta))\right)$$

$$+ \sum_{\{i:y_i=k\}} \log\left(2\exp(x_i^p \gamma) + \frac{\exp(kx_i^\lambda \beta - \exp(x_i^\lambda \beta))}{k!}\right)$$

$$+ \sum_{\{i:y_i \neq 0,k\}} \left(y_i x_i^\lambda \beta - \exp(x_i^\lambda \beta) - \log(y_i!)\right).$$

The maximum likelihood estimate $(\widehat{\gamma}, \widehat{\beta})$ is the solution to the equations

$$\frac{\partial \ell(\gamma, \beta)}{\partial \gamma} = 0, \quad \frac{\partial \ell(\gamma, \beta)}{\partial \beta} = 0.$$

These regression parameters can be solved iteratively using Newton-Raphson. In large samples, the MLEs $(\widehat{\gamma}, \widehat{\beta})$ are approximately normal with means (γ, β). The covariance matrix of the estimates can be obtained by taking the inverse of the negative Hessian matrix calculated using the ML estimates. The standard errors are simply the square root of the diagonal elements of the covariance matrix.

3.2 Grouped Frequency Data

Rather than covariates consisting of subject-specific measurements, we could alternatively face situations where there are only $l = 1, \ldots, s$ $(s < n)$ sets of

Table 5 General layout of grouped data

Response	Covariates		
$n_{01}, n_{11}, \ldots, n_{m1}$	x_{11}	$\ldots$	x_{1u}
$n_{02}, n_{12}, \ldots, n_{m2}$	x_{21}	$\ldots$	x_{2u}
$\vdots$	$\vdots$	$\ddots$	$\vdots$
$n_{0s}, n_{1s}, \ldots, n_{ms}$	x_{s1}	$\ldots$	x_{su}

covariate patterns (e.g., factorial experiments). As shown in Table 5, here our data consists of the frequencies $n_{0l}, n_{1l}, \ldots, n_{ml}$, where n_{jl} represents the frequencies of count j $(j = 0, \ldots, m)$, where m is the largest observed frequency and $n = \sum_{l=1}^{s} \sum_{j=1}^{m} n_{jl}$.

The binomial proportion p_l and Poisson mean λ_l can be parameterized by using the logit and log link functions similar to that provided in (5) as follows:

$$\text{logit}(p_l) = \boldsymbol{x}_l^p \boldsymbol{\gamma} \qquad \text{and} \qquad \log(\lambda_l) = \boldsymbol{x}_l^\lambda \boldsymbol{\beta}, \tag{7}$$

where $\boldsymbol{x}_l^p$ is the set of covariates in the lth covariate pattern used to model p, $\boldsymbol{x}_l^\lambda$ is the set of covariates in the lth covariate pattern used to model λ, and where $\boldsymbol{\gamma}$ and $\boldsymbol{\beta}$ are again the vectors of corresponding regression coefficients. The log-likelihood function then becomes

$$\ell(p_l, \lambda_l | n_{jl}) = \sum_{\{l:j=0\}} n_{jl} \log\left(p_l^2 + q_l^2 \exp(-\lambda_l)\right) + \sum_{\{l:j=k\}} n_{jl} \log\left(2p_l q_l + q_l^2 \frac{\exp(-\lambda_l)\lambda_l^k}{k!}\right)$$

$$+ \sum_{\{l:\,j=1 \atop \neq k\}}^{m} n_{jl} \log\left(q_l^2 \frac{\exp(-\lambda_l)\lambda_l^j}{j!}\right),$$

which can be reexpressed in terms of the regression parameters $\boldsymbol{\gamma}$ and $\boldsymbol{\beta}$ as

$$\ell(\boldsymbol{\gamma}, \boldsymbol{\beta}) = -2 \sum_{l:j=0}^{m} n_{jl} \log\left(1 + \exp(\boldsymbol{x}_l^p \boldsymbol{\gamma})\right)$$

$$+ \sum_{\{l:j=0\}} n_{jl} \log\left(\exp(2\boldsymbol{x}_l^p \boldsymbol{\gamma}) + \exp(-\exp(\boldsymbol{x}_l^\lambda \boldsymbol{\beta}))\right)$$

$$+ \sum_{\{l:j=k\}} n_{jl} \log\left(2\exp(\boldsymbol{x}_l^p \boldsymbol{\gamma}) + \frac{\exp(k\boldsymbol{x}_l^\lambda \boldsymbol{\beta} - \exp(\boldsymbol{x}_l^\lambda \boldsymbol{\beta}))}{k!}\right)$$

$$+ \sum_{\{l:\,j=1 \atop \neq k\}}^{m} n_{jl}\left(j\boldsymbol{x}_l^\lambda \boldsymbol{\beta} - \exp(\boldsymbol{x}_l^\lambda \boldsymbol{\beta}) - \log(j!)\right).$$

The ML estimates $(\widehat{\gamma}, \widehat{\beta})$ are again found as the solution to the first-order derivatives

$$\frac{\partial \ell(\gamma, \beta)}{\partial \gamma} = 0, \ \frac{\partial \ell(\gamma, \beta)}{\partial \beta} = 0.$$

Estimated standard errors of the regression parameters can be found by evaluating the inverse information matrix at the ML estimates.

3.3 Dental Epidemiology Example

We now return to the dental epidemiology data (Bohning et al. 1997, 1999), a subset of which is presented in Table 6. The aim of the study was to compare six methods of school-based dental care: 1. oral health education, 2. enrichment of the school diet with rice bran, 3. mouthwash with 0.2% of NaF solution, 4. oral hygiene, 5. all of the four treatments, and 6. a standard care control. Gender and race/ethnicity groups (White, Black, Others including predominantly Hispanic) were also considered.

Inspection across all 1013 child measures shows inflated counts at 0 and 1. Thus, the $DIP(p, \lambda)$ regression model is used to assess whether the $\delta(DMFT)$ counts were associated with the treatment and covariates, accounting for possible inflation at 0 or 1. For simplicity, both p and λ were modeled against the same set of covariates (treatment, gender, and race/ethnicity). The regression parameter estimates are provided in Table 7. We see that the combination of all four treatments lowers the likelihood of inflated counts relative to the control, Black race/ethnicity leads to a lower likelihood of inflated counts relative to children with Other or Hispanic race/ethnicities, and there was no effect on the probability of inflation due to gender. For the traditional Poisson distribution, we see that the education treatment leads to lower expected counts than the control, and neither gender nor race/ethnicity had a significant relationship with the expected $\delta(DMFT)$ counts.

Table 6 Dental epidemiology data

ID	Treatment	Gender	Ethnicity	δ(DMFT)
1	1	Male	White	0
2	2	Female	White	1
3	4	Female	Black	2
4	5	Female	Other	0
$\vdots$	$\vdots$	$\vdots$	$\vdots$	$\vdots$
1012	3	Male	Black	3
1013	2	Male	Other	0

Table 7 Regression parameter estimates for dental epidemiology data

Parameter	Logit link for p		Log link for λ	
	Est. (S.E.)	p-value	Est. (S.E.)	p-value
Constant	1.484 (0.677)	0.029	0.878 (0.190)	0.001
Treatment				
Educ	1.161 (0.807)	0.151	−0.451 (0.169)	0.008
Enrich	0.847 (0.772)	0.273	−0.254 (0.180)	0.158
Rinse	−0.928 (0.544)	0.088	−0.003 (0.159)	0.983
Hygiene	−0.760 (0.572)	0.185	−0.041 (0.200)	0.838
All	−1.195 (0.543)	0.028	−0.097 (0.167)	0.561
Gender				
Male	0.077 (0.236)	0.774	0.113 (0.085)	0.184
Ethnicity				
White	−0.181 (0.354)	0.609	0.175 (0.109)	0.111
Black	−0.766 (0.362)	0.035	0.060 (0.137)	0.659

4 Conclusion

In this chapter we have introduced a doubly-inflated Poisson (DIP) distribution to model the probabilities of count data that feature inflation at some value k in addition to inflation at zero. Both maximum likelihood and moment estimators were derived, though the latter did not estimate the Poisson mean parameter as efficiently as the likelihood-based estimator. The DIP distribution was also incorporated into the generalized linear models framework, and likelihood-based estimation procedures were developed to estimate the regression coefficients.

A natural extension would be to model the inflation at value k with an additional parameter than that used to model the inflation at zero (as mentioned earlier). Though this parameterization would allow better modeling of both inflationary tendencies, the added complexity could complicate the computations needed for parameter estimation. The trade-off in this instance between more accurate modeling and computational complexity remains to be seen. Other extensions include allowing inflation at three or more values, and incorporating a process to determine whether non-zero inflation exists, as opposed to "knowing" that a certain value exhibits the inflation. Extensions to correlated counts measures are also possible.

Acknowledgements Dr. Manasi Sheth-Chandra's research was partially supported by a Modeling and Simulation graduate student award at Old Dominion University.

Appendix: Analytical Derivatives for DIP(p, λ) Distribution

Information Matrix

Let Y be distributed as $DIP(p, \lambda)$ given in (1). The Fisher information matrix for this distribution is

$$\mathcal{I} = \begin{pmatrix} I(p) & I(p, \lambda) \\ I(p, \lambda) & I(\lambda) \end{pmatrix}, \tag{8}$$

where the elements of $\mathcal{I}$ are defined as

$$I(p) = -E\left(\frac{\partial^2 \log(f(y; p, \lambda))}{\partial p^2}\right)$$

$$= \frac{-2\left(1 + \exp(-\lambda)\right)\left(p^2 + q^2 \exp(-\lambda)\right) - \left(2p - 2q \exp(-\lambda)\right)^2}{\left(p^2 + q^2 \exp(-\lambda)\right)}$$

$$- \left[\frac{\left(2pq + q^2 \frac{\exp(-\lambda)\lambda^k}{k!}\right)\left(-4 + 2\frac{\exp(-\lambda)\lambda^k}{k!}\right) - \left(2 - 4p - 2q\frac{\exp(-\lambda)\lambda^k}{k!}\right)^2}{\left(2pq + q^2 \frac{\exp(-\lambda)\lambda^k}{k!}\right)}\right]$$

$$+ 2\left(1 - \exp(-\lambda) - \frac{\exp(-\lambda)\lambda^k}{k!}\right), \tag{9a}$$

$$I(p, \lambda) = -E\left(\frac{\partial^2 \log(f(y; p, \lambda))}{\partial \lambda \partial p}\right)$$

$$= \frac{-2pq \exp(-\lambda)}{\left(p^2 + q^2 \exp(-\lambda)\right)} - \frac{2q^2 \exp(-\lambda)\lambda^k}{\left(2pq(k!) + q^2 \exp(-\lambda)\lambda^k\right)}\left(1 - \frac{k}{\lambda}\right), \tag{9b}$$

and

$$I(\lambda) = -E\left(\frac{\partial^2 \log(f(y; p, \lambda))}{\partial \lambda^2}\right)$$

$$= \frac{-p^2 q^2 \exp(-\lambda)}{\left(p^2 + q^2 \exp(-\lambda)\right)} + \frac{q^2}{\lambda}\left(1 - \frac{\exp(-\lambda)\lambda^{k-1}}{(k-1)!}\right)$$

$$+ \frac{q^2 \exp(-\lambda)\lambda^k}{\left(2pq(k!) + q^2 \exp(-\lambda)\lambda^k\right)} \times \left(\left(\frac{k}{\lambda} - 1\right)^2 q^2 \frac{\exp(-\lambda)\lambda^k}{k!}\right)$$

$$- q^2 \frac{\exp(-\lambda)\lambda^k}{k!}\left(\left(\frac{k}{\lambda} - 1\right)^2 - \frac{k}{\lambda^2}\right). \tag{9c}$$

The covariance matrix of Y and Y^2 is

$$\Sigma = \begin{pmatrix} Var(Y) & Cov(Y, Y^2) \\ Cov(Y, Y^2) & Var(Y^2) \end{pmatrix}$$

where

$$Var(Y) = 2pqk^2 + q^2(\lambda^2 + \lambda) - \left(2pqk + q^2\lambda\right)^2,$$

$$Var(Y^2) = 2pqk^4 + q^2\left(\lambda^4 + 6\lambda^3 + 7\lambda^2 + \lambda\right) - \left(2pqk^2 + q^2\left(\lambda + \lambda^2\right)\right)^2, \text{ and}$$

$$Cov(Y^2, Y) = Cov(Y, Y^2)$$

$$= \left(2pqk^3 + q^2\left(\lambda^3 + 3\lambda^2 + \lambda\right)\right) - \left((2pqk + q^2\lambda)(2pqk^2 + q^2\left(\lambda + \lambda^2\right))\right).$$

Score Equations for DIP(p, λ) Regression

The partial derivatives for the log-likelihood function for $DIP(p, \lambda)$ regression of raw counts are

$$\frac{\partial \ell(\gamma, \beta)}{\partial \gamma} = -2\sum_{i=1}^{n} \frac{x_i^P \exp(x_i^P \gamma)}{1 + \exp(x_i^P \gamma)} + \sum_{\{i:y_i=0\}} \frac{2x_i^P \exp(2x_i^P \gamma)}{\exp(2x_i^P \gamma) + \exp(-\exp(x_i^\lambda \beta))}$$

$$+ \sum_{\{i:y_i=k\}} \frac{2x_i^P \exp(x_i^P \gamma)}{2\exp(x_i^P \gamma) + \dfrac{\exp(kx_i^\lambda \beta - \exp(x_i^\lambda \beta))}{k!}},$$

and

$$\frac{\partial \ell(\gamma, \beta)}{\partial \beta} = \sum_{\{i:y_i=0\}} \frac{x_i^\lambda \exp(x_i^\lambda \beta - \exp(x_i^\lambda \beta))}{\exp(2x_i^P \gamma) + \exp(-\exp(x_i^\lambda \beta))}$$

$$+ \sum_{\{i:y_i=k\}} \frac{\left(kx_i^\lambda - x_i^\lambda \exp(x_i^\lambda \beta)\right)\exp(kx_i^\lambda \beta - \exp(x_i^\lambda \beta))}{k!\left(2\exp(x_i^P \gamma) + \dfrac{\exp(kx_i^\lambda \beta - \exp(x_i^\lambda \beta))}{k!}\right)}$$

$$+ \sum_{\{i:\substack{y_i=y \\ y\neq 0,k}\}} \left(y_i x_i^\lambda - x_i^\lambda \exp(x_i^\lambda \beta)\right).$$

The partial derivatives for the log-likelihood function for $DIP(p, \lambda)$ regression of grouped frequencies are

$$\frac{\partial \ell(\boldsymbol{\gamma}, \boldsymbol{\beta})}{\partial \boldsymbol{\gamma}} = -2 \sum_{l:j=0}^{m} n_{jl} \frac{x_l^p \exp(x_l^p \boldsymbol{\gamma})}{1 + \exp(x_l^p \boldsymbol{\gamma})} + \sum_{\{l:j=0\}} n_{jl} \frac{2x_l^p \exp(2x_l^p \boldsymbol{\gamma})}{\exp(2x_l^p \boldsymbol{\gamma}) + \exp(-\exp(x_l^\lambda \boldsymbol{\beta}))}$$

$$+ \sum_{\{l:j=k\}} n_{jl} \frac{2x_l^p \exp(x_l^p \boldsymbol{\gamma})}{2 \exp(x_l^p \boldsymbol{\gamma}) + \dfrac{\exp(k x_l^\lambda \boldsymbol{\beta} - \exp(x_l^\lambda \boldsymbol{\beta}))}{k!}},$$

and

$$\frac{\partial \ell(\boldsymbol{\gamma}, \boldsymbol{\beta})}{\partial \boldsymbol{\beta}} = - \sum_{\{l:j=0\}} n_j \frac{x_l^\lambda \exp(x_l^\lambda \boldsymbol{\beta} - \exp(x_l^\lambda \boldsymbol{\beta}))}{\exp(2x_l^p \boldsymbol{\gamma}) + \exp(-\exp(x_l^\lambda \boldsymbol{\beta}))}$$

$$+ \sum_{\{l:j=k\}} n_{jl} \frac{\left(k x_l^\lambda - x_l^\lambda \exp(x_l^\lambda \boldsymbol{\beta})\right) \exp(k x_l^\lambda \boldsymbol{\beta} - \exp(x_l^\lambda \boldsymbol{\beta}))}{k! \left(2 \exp(x_l^p \boldsymbol{\gamma}) + \dfrac{\exp(k x_l^\lambda \boldsymbol{\beta} - \exp(x_l^\lambda \boldsymbol{\beta}))}{k!}\right)}$$

$$+ \sum_{\{l: \substack{j=1 \\ \neq k}\}}^{m} n_{jl} \left(j x_l^\lambda - x_l^\lambda \exp(x_l^\lambda \boldsymbol{\beta})\right).$$

References

Bae, S., Famoye, F., Wulu, J.T., Bartolucci, A.A., Singh, K.P.: A rich family of generalized Poisson regression models with applications. Math. Comput. Simul. **69**, 4–11 (2005)

Böhning, D., Dietz, E., Schlattmann, P.: Zero-inflated count models and their applications in public health and social science. In: Applications of Latent Trait and Latent Class Models in the Social Sciences, pp. 333–344. Wasemann, Münster (1997)

Böhning, D., Dietz, E., Schlattmann, P., Mendonça, L., Kirchner, U.: The zero-inflated Poisson model and the decayed, missing, and filled teeth index in dental epidemiology. J. R. Stat. Soc. **162**, 195–209 (1999)

Cameron, A.C., Trivedi, P.K.: Regression Analysis of Count Data. Cambridge, New York (1998)

Chaganty, N.R., Shi, G.: A note on the estimation of autocorrelation in repeated measurements. Commun. Stat. Theory Methods **33**, 1157–1170 (2004)

Cohen, A.C.: Estimation in mixtures of discrete distributions. In: Proceedings of the International Symposium on Discrete Distributions, Montreal, pp. 373–378. Statistical Pub. Society, Calcutta (1963)

Coxe, S., West, S., Aiken, L.S.: The analysis of count data: a gentle introduction to Poisson regression and its alternative. J. Pers. Assess. **91**, 121–136 (2009)

Hall, D.: Zero-inflated Poisson and binomial regression with random effects: a case study. Biometrics **56**, 1030–1039 (2000)

Hall, D., Shen, J.: Robust estimation for zero-inflated Poisson regression. Scand. J. Stat. **37**, 237–252 (2010)

Johnson, N.L., Kotz, S.: Distributions in Statistics: Discrete Distributions. Houghton Mifflin, Boston (1969)

Lambert, D.: Zero-inflated Poisson regression with an application to defects in manufacturing. Technometrics **34**, 1–14 (1992)

Lin, T.H., Tsai, M.-H.: Modeling health survey data with excessive zero and K responses. Stat. Med. **32**, 1572–1583 (2013)

Marazzi, A., Paccaud, F., Ruffieux, C., Beguin, C.: Fitting the distributions of length of stay by parametric models. Med. Care **36**, 915–927 (1998)

McCulloch, C.E., Searle, S.R.: Generalized, Linear, and Mixed Models. Wiley, New York (2001)

Sheth-Chandra, M.: The doubly inflated Poisson and related regression models. PhD Dissertation. Old Dominion University, Norfolk (2011)

Multivariate Doubly-Inflated Negative Binomial Distribution Using Gaussian Copula

Joseph Mathews, Sumen Sen, and Ishapathik Das

1 Introduction

Multivariate count data appear in many fields of study such as econometrics, ecology, criminology, and epidemiology. One popular alternative to the Poisson distribution for modeling such multivariate count data is the negative binomial distribution. Unlike the Poisson distribution, the negative binomial distribution has a separate mean and variance parameter making it an attractive choice for overdispersed or underdispersed data. In the case of multivariate, zero-inflated data, both the zero-inflated Poisson (ZIP) distribution model and zero-inflated negative binomial (ZINB) distribution model are used. In certain cases, the multivariate count data contain an additional inflation point. The ZIP and ZINB models are then extended to a doubly or multiple inflation model. Forms of such doubly or multiple inflation models using the multivariate Poisson distribution are given by Karlis and Ismail (2005), Sen et al. (2017), and Sengupta et al. (2015). Bivariate zero-inflated negative binomial models exist in the literature such as So et al. (2011) and Ismail and Faroughi (2017). However, little is found on a doubly or multiple inflation model using the multivariate negative binomial distribution.

J. Mathews · S. Sen (✉)
Department of Mathematics and Statistics, Austin Peay State University, Clarksville, TN, USA
e-mail: jmathews6@my.apsu.edu; sens@apsu.edu

I. Das
Department of Mathematics, Indian Institute of Technology Tirupati, Tirupati, India
e-mail: ishapathik@iittp.ac.in

© Springer Nature Switzerland AG 2019

147

N. Diawara (ed.), *Modern Statistical Methods for Spatial and Multivariate Data*,
STEAM-H: Science, Technology, Engineering, Agriculture, Mathematics & Health,
https://doi.org/10.1007/978-3-030-11431-2_8

Expressions of the multivariate negative binomial distribution exist in the literature such as Doss (1979) and Antzoulakos and Philippou (1991). However, such forms are given by the probability generating function and are very complicated in practice. Therefore, copula methods for modeling multivariate count data are a common alternative (Karlis and Nikoloulopoulos 2009; Brechmann et al. 2011). A copula is a probability function commonly used to form the joint probability distribution of $X_1, X_2, \ldots, X_p$ random variables with marginal distributions $F_1, F_2, \ldots, F_p$, respectively. Moreover, a copula preserves the dependence structure between the p random variables in a tractable form. Here, we use copula methods to obtain a multivariate form of a doubly-inflated negative binomial distribution.

This paper is organized as follows. Section 2 will cover the definition of a copula and the fundamental Sklar's theorem (Sklar 1959). Section 3 will cover multivariate discrete distributions using copula methods and Sklar's theorem. In Sect. 4 we construct and provide properties of the doubly-inflated negative binomial model and end with a description of a data simulation algorithm. In Sect. 5 we review maximum likelihood estimation, provide an expression for the likelihood function of the proposed model, and outline a method for estimating the model's parameters. Also, we give parameter and mean squared error (MSE) results for simulated data. In Sect. 6 we apply the model to the DoctorAUS dataset from the *Ecdat* package in R (Bolker 2016) and provide parameter estimates along with their respective asymptotic standard error. We end with a conclusion in Sect. 7.

2   The Multivariate Copula and Sklar's Theorem

A copula is a multivariate distribution function with uniform margins on the interval [0, 1]. Given specified marginal distributions, one can obtain a multivariate probability distribution through a coupla, see Joe (2014) and Song (2007).

Definition 1 A p-dimensional copula is a function $C : [0, 1]^p \rightarrow [0, 1]$ with the following properties:

1. $C(1, \ldots, a_i, \ldots, 1) = a_i, \forall i = 1, 2, \ldots, p$ and $a_i \in [0, 1]$
2. $C(a_1, a_2, \ldots, a_p) = 0$ if at least one $a_i = 0$ for $i = 1, 2, \ldots, p$.
3. For any $a_{i,1}, a_{i,2} \in [0, 1]$ with $a_{i,1} \leq a_{i,2}$, for $i = 1, 2, \ldots, p$:

$$\sum_{j_1=1}^{2} \sum_{j_2=1}^{2} \cdots \sum_{j_p=1}^{2} (-1)^{j_1+j_2+\cdots+j_p} C(a_{1,j_1}, a_{2,j_2}, \ldots, a_{n,j_p}) \geq 0.$$

Using Sklar's theorem, one can form a multivariate probability distribution function using known univariate marginal distributions and a unique copula.

Theorem 1 (Sklar's Theorem) *Let* $X_1, X_2, \ldots, X_p$ *be random variables with marginal distribution functions* $F_1, F_2, \ldots, F_p$ *and joint cumulative distribution function* F, *then the following holds:*

1. *There exists a p-dimensional copula C such that for all* $x_1, x_2, \ldots, x_p \in \mathbb{R}$

$$F(x_1, x_2, \ldots, x_p) = C(F_1(x_1), F_2(x_2), \ldots, F_p(x_p))$$

2. *If* $X_1, X_2, \ldots, X_p$ *are continuous, then the copula C is unique. Otherwise, C can be uniquely determined on a p-dimensional rectangle with dimension* $Range(F_1) \times Range(F_2) \times \ldots \times Range(F_p)$.

See Joe (2014) and Nelsen (2006) for more information of copula's and associated properties.

One very common copula is the Gaussian copula due to its ability to preserve the correlation structure of two variables.

Definition 2 The Gaussian copula is given by the function:

$$C(u_1, u_2, \ldots, u_p | R(r)) = \Phi_R(\Phi^{-1}(u_1), \Phi^{-1}(u_2), \ldots, \Phi^{-1}(u_p)), \qquad (1)$$

where Φ^{-1} is the inverse CDF of a standard normal and Φ_R is the joint cumulative distribution function of a standard multivariate normal distribution with covariance matrix equal to the correlation matrix R.

Definition 3 The Gaussian copula density is defined as:

$$c(u_1, u_2, \ldots, u_p | R(r)) = \frac{1}{\sqrt{|R(r)|}} exp(-\frac{1}{2} U^T \times (R(r)^{-1} - I_p) \times U), \qquad (2)$$

where $U = (\Phi^{-1}(u_1), \Phi^{-1}(u_2), \ldots, \Phi^{-1}(u_p))^T$

A plot of a bivariate Gaussian copula with $r = 0.85$ is given in Fig. 1. If the dimension of the random variable X is p, then there are $\binom{p}{2}$ elements in the association matrix R. As p increases, the number of parameters increases. If a model contains too many parameters to estimate, then computational challenges may be experienced. Therefore, we assume a structured correlation matrix R of two kinds.

1. Equi-correlation structure: Under this structure, we assume

$$R(r) = r\mathbf{1}\mathbf{1}^t - (1 - r)I_p$$

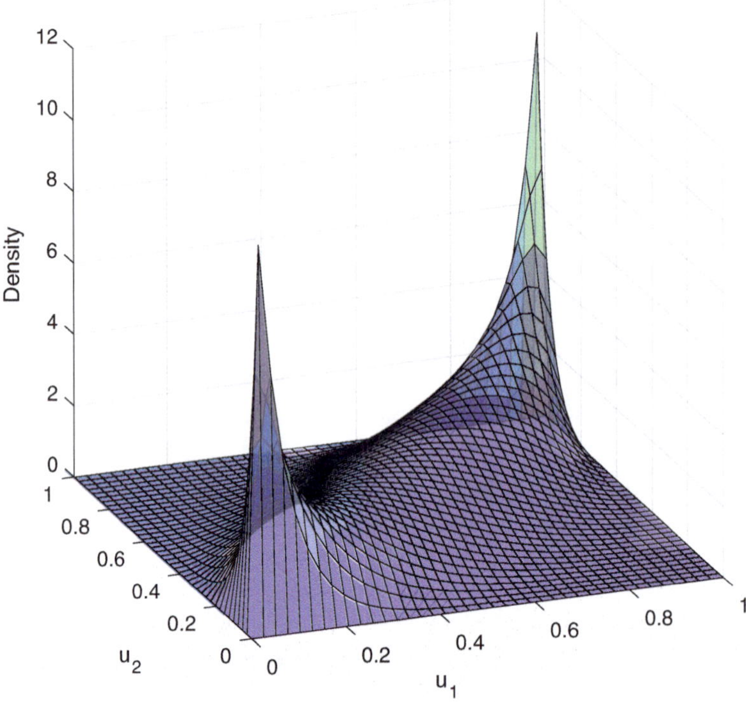

Fig. 1 A bivariate Gaussian copula density with $r = 0.85$

where I_p is a p-dimensional identity matrix, $r \in (-\frac{1}{p-1}, 1)$, and $\mathbf{1}$ is a p-dimensional column vector of ones. It follows from Olkin and Pratt (1958) that:

$$R^{-1}(r) = \frac{1}{1-r} I_p - \frac{r}{(1-r)\{1 + (p-1)r\}} \mathbf{1}\mathbf{1}^t \tag{3}$$

2. AR-1 structure: Under this structure, the (i, j)th element of $R(r)$ is given by $r^{|i-j|}$, with $r \in (-1, 1)$. The inverse of this matrix is given below (Chaganty 1997)

$$R^{-1}(r) = \frac{1}{1-r^2} (I_p - r^2 M_2 - r M_1), \tag{4}$$

where $M_2 = diag(0, 1, \ldots, 1, 0)$ and M_1 is a tridiagonal matrix with 0 on the main diagonal and 1 on the upper and lower diagonals.

3 Discrete Multivariate Density Function Using the Gaussian Copula

A copula may also be used to derive a joint distribution for discrete data. Given a set of discrete, marginal distributions $F_1(y_1|\theta_1)$, $F_2(y_2|\theta_2), \ldots, F_p(y_p|\theta_p)$, one can obtain the following joint probability mass function of $\mathbf{Y} = (Y_1, \ldots, Y_p)$:

$$f_{C_\Phi}(\mathbf{Y}|\Theta, R(r)) = P(Y_1 = y_1, Y_2 = y_2, \ldots, Y_p = y_p|\Theta, R(r))$$
$$= \sum_{j_1=1}^{2} \sum_{j_2=1}^{2} \cdots \sum_{j_p=1}^{2} (-1)^{j_1+j_2+\cdots+j_p} C_\Phi(u_{1,j_1}, \ldots, u_{p,j_p}|R(r))$$

$$(5)$$

where $\Theta = (\theta_1, \ldots, \theta_p)$, $C_\Phi(\cdot|R(r))$ is the Gaussian copula distribution function with association matrix $R(r)$. Note that $u_{j1} = F_j(y_j)$ and $u_{j2} = F_j(y_j-)$, where $F_j(y_j-)$ is the left-hand limit of F_j at y_j equal to $F_j(y_j - 1)$.

Let $Y_1, \ldots, Y_n$ be n random variables where the ith random variable Y_i has a negative binomial distribution with probability and size parameters p_i and s_i, respectively. The distribution function be $F_{s_i, p_i}(y_i)$ and probability mass function given by:

$$f_{NB}(y_i|s_i, p_i) = P(Y_i = y_i) = \binom{y_i + s_i - 1}{y_i}(1 - p_i)^{s_i} p_i^{y_i}, \qquad (6)$$

where $s_i > 0$, $y_i \in \mathbb{N}$ and $p_i, \in [0, 1]$ for $i = 1, \ldots n$. Using Sklar's theorem and Eq. (5) one can derive the joint probability mass function $f_{C_\Phi}(Y_1 = y_1, \ldots Y_n = y_n)$. For bivariate case plots of the joint probability mass function are given in Fig. 2 with different correlation values.

4 Multivariate Doubly-Inflated Negative Binomial Model

Modeling count data are very popular in statistics. Both the Poisson and negative binomial distributions are very popular to model count data. One simple approach of introducing correlation among count variables is through common additive error models. Kocherlakota and Kocherlakota (2001) and Johnson et al. (1997) provided a detailed discussion for the one-factor multivariate Poisson model. Along this line of study, Winkelmann (2000) proposed a multivariate negative binomial regression model. In the case of zero and doubly-inflated count data, zero-inflated and doubly-inflated multivariate Poisson models are available (Sen et al. 2017; Agarwal et al. 2002; Lee et al. 2009). In this section, we shortly review the model construction process. For details, refer to Sen et al. (2017).

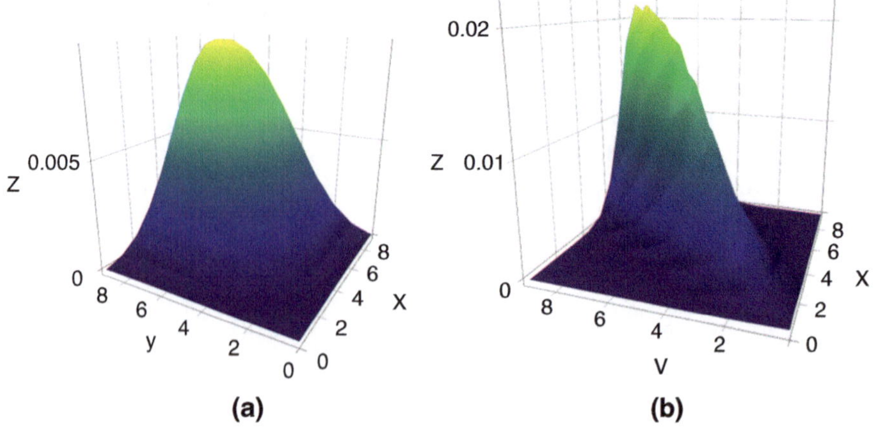

Fig. 2 Bivariate negative binomial density using a Gaussian copula with $s_1 = 10$, $p_1 = 0.50$ and $s_2 = 15$, $p_2 = 0.60$. (a) $r = 0.10$. (b) $r = 0.90$

4.1 Construction of Doubly-Inflated Model

Consider a latent random variable Z with the following probability distribution:

$$P(Z = z) = \begin{cases} p_{r_1} & \text{if } z = 2 \\ p_{r_2} & \text{if } z = 1 \\ 1 - q & \text{if } z = 0 \\ 0 & \text{elsewhere} \end{cases} \tag{7}$$

where $p_{r_1}, p_{r_2} \in (0, 1)$ and $q = p_{r_1} + p_{r_2} < 1$. Also, let $\mathbf{Y} = (Y_1, \ldots, Y_p)$ be a multivariate negative binomial random variable with mass function f_{C_Φ} constructed using a Gaussian copula as mentioned in Eq. (5).

The distribution functions of $\mathbf{Y}|Z$ and $(\mathbf{Y}, Z)$ are given below:

$$f_1(\mathbf{Y}|Z) = \begin{cases} 1 & \text{if } z = 2, y = (0, \ldots, 0) \\ 1 & \text{if } z = 1, y = (k_1, \ldots, k_p) \\ f_{C_\Phi}(y_1, \ldots, y_p) & \text{if } z = 0, y = (y_1, \ldots, y_p) \end{cases} \tag{8}$$

$$f_2(\mathbf{Y}, Z) = \begin{cases} p_{r_1} & \text{if } z = 2, y = (0, \ldots, 0) \\ p_{r_2} & \text{if } z = 1, y = (k_1, \ldots, k_p) \\ (1 - q)f_{C_\Phi}(y_1, \ldots, y_p) & \text{if } z = 0, y = (y_1, \ldots, y_p) \end{cases} \tag{9}$$

Thus, the probability mass function of the multivariate doubly-inflated negative binomial distribution is given by :

$$
f_{R(r)}(\mathbf{Y}|\Theta) = \begin{cases} p_{r_1} + (1-q)f_{C_\Phi}(0, \dots, 0|R(\mathbf{r})) & \text{if } y = (0, \dots, 0) \\ p_{r_2} + (1-q)f_{C_\Phi}(k_1, \dots, k_p|R(\mathbf{r})) & \text{if } y = (k_1, \dots, k_p) \\ (1-q)f_{C_\Phi}(y_1, \dots, y_p|R(\mathbf{r})) & \text{if } y = (y_1, \dots, y_p), \end{cases} \quad (10)
$$

where $\Theta = (p_{r_1}, p_{r_2}, p_1, \dots, p_p, s_1, \dots s_p)$. The parameters p_{r_1} and p_{r_2} correspond to the points of inflation in the cells $(0, \dots, 0)$ and $(k_1, \dots, k_p)$, respectively.

4.2 *Bivariate Doubly-Inflated Negative Binomial*

In the previous section, a multivariate doubly-inflated model is developed using a Gaussian copula. In order to simplify things we consider a bivariate doubly-inflated negative binomial model with correlation structure $R(r) = \left(\begin{smallmatrix} 1 & r \\ r & 1 \end{smallmatrix} \right)$. For bivariate case consider the bivariate random vector $\mathbf{Y} = (Y_1, Y_2)$. Using Eq. (10), the joint mass function of $\mathbf{Y}$ is given below.

First consider the case when $\mathbf{Y} = (0, 0)$, then Eq. (5) reduces to:

$$
\begin{aligned}
f_{C_\Phi}(\mathbf{Y}|\Theta, R(r)) &= P(Y_1 = 0, Y_2 = 0) \\
&= C_\Phi(F_{s_1,p_1}(0), F_{s_2,p_2}(0)) - C_\Phi(F_{s_1,p_1}(-1), F_{s_2,p_2}(0)) \\
&\quad - C_\Phi(F_{s_1,p_1}(0), F_{s_2,p_2}(-1)) + C_\Phi(F_{s_1,p_1}(-1), F_{s_2,p_2}(-1)) \\
&= C_\Phi(F_{s_1,p_1}(0), F_{s_2,p_2}(0)) \\
&= C_\Phi((1-p_1)^{s_1}, (1-p_2)^{s_2}).
\end{aligned}
$$

$$(11)$$

This follows from the fact that $F_{s_1,p_1}(-1) = F_{s_2,p_2}(-1) = 0$ and the definition of a copula. Using Eq. (10), the mass function of the bivariate doubly-inflated negative binomial is given by:

$$
f_{R(r)}(\mathbf{Y}|\Theta) = \begin{cases} p_{r_1} + (1-q)C_\Phi((1-p_1)^{s_1}, (1-p_2)^{s_2}), & y = (0, 0) \\ p_{r_2} + (1-q)f_{C_\Phi}(k_1, k_2|R(r)), & y = (k_1, k_2) \\ (1-q)f_{C_\Phi}(y_1, y_2|R(r)), & otherwise. \end{cases} \quad (12)
$$

A plot of Eq. (12) is given in Fig. 3.

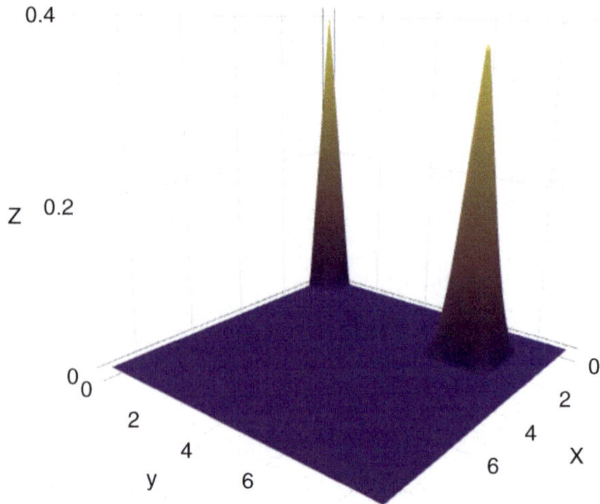

Fig. 3 Bivariate doubly-inflated negative binomial model with $s_1 = 10, s_2 = 15, p_1 = 0.50, p_2 = 0.60$ and inflation points at $(0, 0)$ and $(7, 2)$

4.3 Properties of Multivariate Doubly-Inflated Negative Binomial

In this section we discuss important properties of the proposed distribution in Eq. (10). Expression for marginal mass functions and nth order moments are provided.

Marginal Distributions

From the multivariate distribution mentioned in Eq. (10), marginal distribution of Y_i is given by:

$$
m_i(y_i) = \begin{cases} p_{r_1} + (1 - q)(1 - p_i)^{s_i}, & y_i = 0 \\ p_{r_2} + (1 - q) f_{NB}(k_i), & y_i = k_i \\ (1 - q) f_{NB}(y_i), & y_i = \mathbb{N} - \{\{0\}, \{k_i\}\}, \end{cases} \tag{13}
$$

where $f_{NB)}(.)$ is the mass function of a negative binomial distribution, with parameters p_1 and s_1, defined in Eq. (6). Plot of the mass function is provided in Fig. 4.

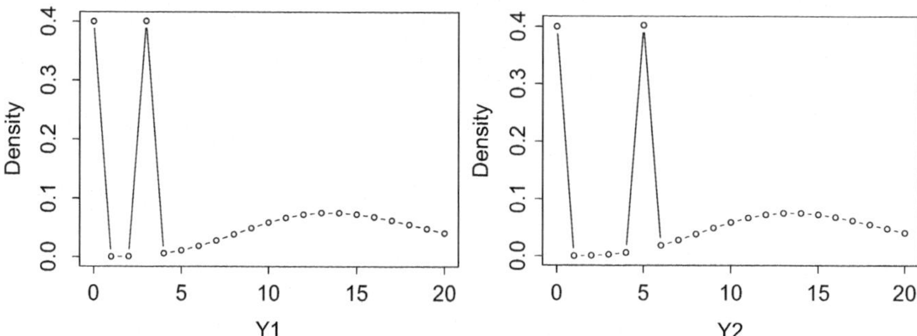

Fig. 4 Mass functions of Y_1 and Y_2 where $k_1 = 3$ and $k_2 = 5$

Using Eqs. (10) and (13), the expected value of the distribution can be computed as below:

$$E(Y_i) = \sum_{y_1=0}^{\infty} \cdots \sum_{y_i=0}^{\infty} \cdots \sum_{y_p=0}^{\infty} y_i f_{R(r)}(y_1, \ldots, y_p | \Theta)$$

$$= \sum_{y_i=0}^{\infty} y_i \sum_{y_1=0}^{\infty} \cdots \sum_{y_{i-1}=0}^{\infty} \sum_{y_{i+1}=0}^{\infty} \cdots \sum_{y_p=0}^{\infty} f_{R(r)}(y_1, \ldots, y_p | \Theta)$$

$$= \sum_{y_i=0}^{\infty} y_i m_i(y_i)$$

$$= 0 m_i(0) + k_i m_i(k_i) + \sum_{\substack{x=1 \\ x \neq 0, k_i}}^{\infty} y_i m(y_i)$$

$$= k_i \{ p_{r_2} + (1-q) f_{NB}(k_i) \} + \sum_{\substack{x=1 \\ x \neq 0, k_i}}^{\infty} y_i \{ (1-q) f_{NB}(y_i) \}$$

$$= k_i p_{r_2} + (1-q) \sum_{y_i=0}^{\infty} y_i f_{NB}(y_i)$$

$$= k_i p_{r_2} + (1-q) \frac{p_i s_i}{1 - p_i} \tag{14}$$

In a similar fashion, we can find an expression for $E(Y_i^n)$:

$$E(Y_i^n) = k_i^n p_{r_2} + (1-q) E(N_i^n), \tag{15}$$

where N_i is the negative binomial random variable with parameters p_i and s_i. Using Eqs. (14) and (15), we can find an expression for the variance:

$$V(Y_i) = k_i^2 p_{r_2} + (1 - q)\frac{p_i s_i(1 + p_i s_i)}{(1 - p_i)^2} \\ - [k_i p_{r_2} + (1 - q)(\frac{p_i s_i}{1 - p_i})]^2 \tag{16}$$

4.4   Data Simulation Algorithm

A description of a data simulation algorithm for simulating data from a multivariate distributions using Gaussian copula can be found in Joe (2014) and Nelsen (2006). We are interested in simulating data from the multivariate doubly-inflated negative binomial model given in Eq. (10). The steps for the algorithm are as follows:

1. Simulate Z from the categorical distribution given in Eq. (8)
2. If $Z = 2$, then $y = (0, \ldots, 0)$.
3. If $Z = 1$, then $y = (k, \ldots, k)$.
4. If $Z = 0$, then use the algorithm described in Joe (2014) to simulate n observations from the multivariate mass function defined in Eq. (5).

In particular, we used this algorithm to simulate from bivariate doubly-inflated negative binomial distribution. Below are histograms of the marginals of a simulation with $n = 1000$ samples (Fig. 5).

5   Parameter Estimation Using Maximum Likelihood Estimation

Now we will review and use maximum likelihood techniques to estimate the parameters of our model.

5.1   The Likelihood Function

The doubly-inflated negative binomial probability mass function defined in (14) is dependent on $2 + 2p + \binom{p}{2}$ parameters. Notice that the number of parameters to estimate increases quickly as the number of dimensions increases—posing potential computational problems for high-dimensional data. Thus, we use a structured correlation matrix as defined in Eqs. (3) and (4) to avoid these computational problems. We will use the method of maximum likelihood to estimate the $2 +$

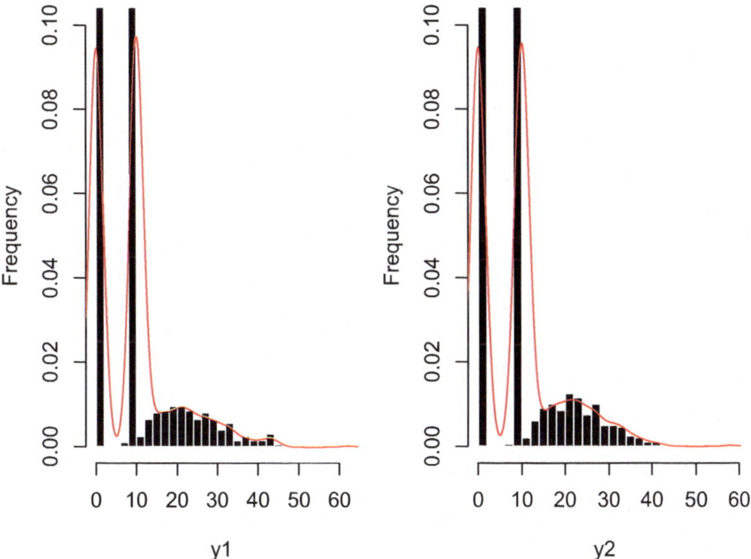

Fig. 5 Histogram of marginal distributions for a bivariate doubly-inflated negative binomial model with $s_1 = 10$, $s_2 = 15$, $p_1 = 0.30$, $p_2 = 0.40$ and inflation point at $k = 10$

$2p + \binom{p}{2}$ parameters. Hence, by (14) the likelihood function for the doubly-inflated negative binomial using a Gaussian copula is given by:

$$L_{R(r)}(\Theta|Y) = \prod_{i:Y_i=(0,\dots,0)} p_{r1} + (1-q)C_\Phi((1-p_1)^{s_1}, \dots, (1-p_p)^{s_p}|R(r))$$

$$\times \prod_{i:Y_i=(k,\dots,k)} p_{r2} + (1-q)f_{C_\Phi}(k, \dots, k|R(r))$$

$$\times \prod_{i:Y_i=(y_{1_i},\dots,y_{n_i})} (1-q)f_{C_\Phi}(y_{1_i}, \dots, y_{n_i}|R(\mathbf{r})) \tag{17}$$

The maximum likelihood parameter estimates are given by the derivative of the log of the likelihood equation in (17). Applying the logarithm function to both sides of (17), we obtain the following:

$$l_{R(r)}(\Theta|Y) = \sum_{i:Y_i=(0,\dots,0)} p_{r1} + (1-q)C_\Phi((1-p_1)^{s_1}, \dots, (1-p_p)^{s_p}|R(r))$$

$$+ \sum_{i:Y_i=(k,\dots,k)} p_{r2} + (1-q)f_{C_\Phi}(k, \dots, k|R(r))$$

$$+ \sum_{i:Y_i=(y_{1_i},\dots,y_{n_i})} (1-q)f_{C_\Phi}(y_{1_i}, \dots, y_{n_i}|R(\mathbf{r})) \tag{18}$$

A complicated expression such as (18) will have no closed form expressions for estimators by solving the score equations. Therefore, we use quasi Newton Raphson (Davidon 1991) to obtain estimates. Also, as estimating parameters from a multivariate Gaussian copula can be computationally challenging, convergence issues arise through direct maximization of the likelihood function. Thus, we will split the estimation process into several steps. The inference function for margins (IFM) method proposed by Joe (2005) is a popular method for estimating copula parameters.

Notice that the algorithm below maximizes the likelihood function in two steps:

1. Start with initial values $p_{r1}{}^0$ and $p_{r2}{}^0$. These are found by finding the proportion of inflation points $(0, \ldots, 0)$ and $(k_1, \ldots, k_p)$ in the data, respectively.
2. Obtain initial estimates for $\mathbf{s}^i = (s_1^i, \ldots, s_p^i)$ and $\mathbf{p}^i = (p_1^i, \ldots, p_p^i)$ by applying the method of maximum likelihood to the marginal distributions of the data.
3. At the ith step, $i = 1, 2, \ldots$, use the initial values from step 2 and maximize the log-likelihood function $l_{R(r)}(p_{r1}{}^0, p_{r2}{}^0, \mathbf{s}^i, \mathbf{p}^i, r)$ with respect to r and obtain $\hat{r}$.
4. Using $\hat{r}$ from step 3, maximize the log-likelihood function $l_{R(r)}(p_{r1}{}^0, p_{r2}{}^0, \mathbf{s}^i, \mathbf{p}^i, \hat{r})$ with respect to $p_{r1}, p_{r2}, \mathbf{s}$, and $\mathbf{p}$ to obtain $\hat{p_{r1}}, \hat{p_{r2}}, \hat{\mathbf{s}}, \hat{\mathbf{p}}$.
5. Index $i = i + 1$ and use the estimates $\hat{p_{r1}}, \hat{p_{r2}}, \hat{\mathbf{s}}$, and $\hat{\mathbf{p}}$ as initial values. Repeat steps 2 and 3 to get updated values.
6. Repeat steps 2, 3, and 4 until convergence.

5.2   Simulation Results

Now the proposed algorithm is used to estimate the doubly-inflated negative binomial model parameters. First, the data simulation algorithm from Sect. 4.2 is used to simulate data from a bivariate, doubly-inflated negative binomial model. The estimation algorithm from Sect. 5.1 is then used to estimate the seven parameters. The results for sample sizes 200 and 500 are given in Table 1.

Table 1 Bivariate, doubly-inflated negative binomial distribution with points of inflation $(0, 0)$ and $(3, 2)$

	Simulation results			
	Sample size = 200		Sample size = 500	
Parameters	Estimates	MSE	Estimates	MSE
$p_1 = 0.4$	0.4129	0.2496	0.4167	0.1217
$p_2 = 0.2$	0.2126	0.0658	0.2140	0.3217
$s_1 = 5$	5.6621	4.3250	5.4683	3.2377
$s_2 = 3$	3.2342	1.0639	3.2339	0.7849
$p_{r1} = 0.30$	0.2967	0.0350	0.3022	0.0200
$p_{r2} = 0.40$	0.3972	0.0647	0.4170	0.0040
$\rho = 0.6$	0.6396	0.2054	0.5591	0.0760

6   Application

We now fit a bivariate doubly-inflated negative binomial distribution to the DoctorAUS dataset. We compare three models: an independent model, a zero-inflated negative binomial (ZINB) model, and a bivariate doubly-inflated negative binomial (BDINB) model. For the independent model, we assume that both variables are independent and a negative binomial distribution is fit to both variables. For the ZINB model, we assume that both variables are independent and a zero-inflated negative binomial distribution is fit to both variables using the *dzinbinom* function from the emdbook package Bolker (2016) in R. For the BDINB model, we fit a doubly-inflated negative binomial distribution and impose a dependence structure on the two variables using Gaussian copula methods. We use Akaike's information criterion (AIC) and the Bayesian information criterion (BIC) to compare the models.

6.1   *Application to DoctorAUS Dataset*

The DoctorAUS dataset can be found in the *Ectdat* package in R. The data comes from a study done at an Australian hospital from 1977 to 1978 with $n = 5190$ observations. We consider the *actdays* $= (0, \ldots, 14)$ and *illness* $= (0, \ldots, 5)$ variables. Here, *actdays* is the number of days of reduced activity from illness or injury in the previous 2 weeks for a given patient and *illness* is the number of days a given patient was sick in the previous 2 weeks. A table of counts for the two variables is given in Table 2. Hence, there appears to be inflation points at $(0, 0)$ and $(1, 0)$. In order to compute the asymptotic standard errors, we estimate p_{r1} and p_{r2} by finding the proportion of $(0, 0)$ and $(1, 0)$ counts rather than maximum likelihood estimation. The results for the parameter estimates are given in Table 3. The results for the three models are given in Table 4.

Table 2 Counts of *actdays* × *illness* variables

Count table															
	0	1	2	3	4	5	6	7	8	9	10	11	12	13	14
0	1543	5	0	0	1	0	0	0	0	0	0	0	0	0	5
1	1375	71	40	26	20	10	6	8	5	2	5	1	2	0	67
2	763	53	28	15	7	18	3	11	5	2	2	1	3	0	35
3	420	25	19	14	8	5	5	10	4	1	1	0	0	2	28
4	200	11	10	8	6	6	1	5	2	1	1	0	0	2	21
5	153	12	11	11	3	1	2	4	1	1	3	0	1	1	32

Table 3 Estimation results for DoctorAUS dataset

Estimation results		
Parameters	Estimates	Standard error
s_1	13.37164495	1.442319086
s_2	0.11853352	0.005808301
p_1	0.85182789	0.013635534
p_2	0.06818760	0.004735758
r	0.03349333	0.031864960

Table 4 Model results for DoctorAUS dataset

Model results		
Model	AIC	BIC
Independent model	24,819.30	24,845.52
ZINB model	27,423.81	27,463.13
BDINB model	245,85.85	25,631.74

7 Conclusion

We have proposed a doubly-inflated model using the multivariate negative binomial distribution. Copula methods were used to preserve the dependence structure of the model in a tractable form. Maximum likelihood estimation was introduced and a method for estimating the proposed model's parameters was given. The doubly-inflated model obtained the best fit compared to the independent and zero-inflated model when applied to real data. For high-dimensional data, computational problems may still occur. Further research includes extending the model to a regression setting by including covariates.

References

Agarwal, D.K., Gelfand, A.E., Citron-Pousty, S.: Zero-inflated models with application to spatial count data. Environ. Ecol. Stat. **9**(4), 341–355 (2002)

Antzoulakos, D.L., Philippou, A.N.: A note on the multivariate negative binomial distributions of order k. Commun. Stat. Theory Methods **20**, 1389–1399 (1991)

Bolker, B.: emdbook: Ecological Models and Data in R; R package version 1.3.9 (2016)

Brechmann, E.C., Czado, C., Kastenmeier, R., Min, A.: A mixed copula model for insurance claims and claim sizes. Scand. Actuar. J. **2012**(4), 278–305 (2011)

Chaganty, N. Rao.: An alternative approach to the analysis of longitudinal data via generalized estimating equations. J. Stat. Planning Inference **63**, 39–54 (1997)

Davidon, W.C.: Variable metric method for minimization. SIAM J. Optim. **1**(1), 1–17 (1991)

Doss, D.C.: Definition and characterization of multivariate negative binomial distribution. J. Multivar. Anal. **9**(3), 460–464 (1979)

Ismail, N., Faroughi, P.: Bivariate zero-inflated negative binomial regression model with applications. J. Stat. Comput. Simul. **87**(3), 457–477 (2017)

Joe, H.: Asymptotic efficiency of the two-stage estimation method for copula-based models. J. Multivar. Anal. **94**, 401–419 (2005)

Joe, H.: Dependence Modeling with Copulas. Chapman & Hall/CRC Monographs on Statistics & Applied Probability. Chapman and Hall/CRC, Boca Raton (2014)

Johnson, N.L., Kotz, S., Balakrishnan, N.: Discrete Multivariate Distributions. Wiley, New York (1997)

Karlis, D., Ismail, N.: Bivariate Poisson and diagonal inflated bivariate Poisson regression models in R. J. Stat. Softw. **14**(10) (2005)

Karlis, D., Nikoloulopoulos, A.: Modeling multivariate count data using copulas. Commun. Stat. Simul. Comput. **39**(1), 172–187 (2009)

Kocherlakota, S., Kocherlakota, K.: Regression in the bivariate Poisson distribution. Commun. Stat. **30**(5), 815–825 (2001)

Lee, J., Jung, B.C., Jin, S.H.: Tests for zero inflation in a bivariate zero-inflated Poisson model. Statistica Neerlandica **63**, 400–417 (2009)

Nelsen, R.B.: An Introduction to Copulas. Springer Series in Statistics. Springer, Berlin (2006)

Olkin, I., Pratt, J.W.: Unbiased estimation of certain correlation coefficients. Ann. Math. Stat. **29**(1), 201–211 (1958)

Sen, S., Sengupta, P., Diawara, N.: Doubly inflated Poisson model using Gaussian copula. Commun. Stat. Theory Methods **10**(0), 1–11 (2017)

Sengupta, Pooja, Chaganty, N.R., Sabo, Roy T.: Bivariate doubly inflated Poisson models with applications. J. Stat. Theory Pract. **10**(1), 202–215 (2015)

Sklar, A.: Fonctions de répartition à n dimensionset leurs marges. Publ. Inst. Statis. Univ. Paris **8**, 229–231 (1959)

So, S., Lee, D., Jung, B.C.: An alternative bivariate zero-inflated negative binomial regression model using a copula. Econ. Lett. **113**(2), 183–185 (2011)

Song, P.X.K.: Correlated Data Analysis: Modeling, Analytics, and Applications, 1st edn., vol. 365. Springer Series in Statistics. Springer, New York (2007)

Winkelmann, R.: Seemingly unrelated negative binomial regression. Oxford Bull. Econ. Stat. **62**(4), 553–560 (2000)

Quantifying Spatio-Temporal Characteristics via Moran's Statistics

Jennifer L. Matthews, Norou Diawara, and Lance A. Waller

1 Introduction

Moran's index is a measure of spatial autocorrelation within a domain area; it has and continues to be applied in many fields. As described by Moran (1950), when given a set of spatial variates (defined on a two-dimensional discrete area), we may want to investigate whether there is any evidence that spatial autocorrelation is present overall or in neighboring clusters based on selected features. For example, in epidemiology, we may be interested in disease mapping based on the number of people infected in a given area. Lawson (2009) and Zhou and Lawson (2008) show evidence of clustered disease maps and develop spatio-temporal disease surveillance under a Bayesian modeling. Meddens and Hicke (2014) describe spatial and temporal patterns of tree mortality from beetle ecology and dynamics in parts of Colorado and Wyoming. The spatial point process experiment in Vaillant et al. (2011) describes sustainability or occurrence dates of sugarcane viruses. In environmental setting, growth of species can be modeled individually, but it may be much more appropriate to include their spatio-temporal dynamics in the ecosystem (Jones-Todd et al. 2018). In geoscience, patterns have always been of interest. Wang et al. (2015) proposed geographically weighted statistics as an alternative to the assumption of homogeneous spatial characteristics, but not adding a time factor.

J. L. Matthews · N. Diawara (✉)
Department of Mathematics and Statistics, Old Dominion University, Norfolk, VA, USA
e-mail: ndiawara@odu.edu

L. A. Waller
Department of Biostatistics, Rollins School of Public Health, Emory University,
Atlanta, GA, USA

© Springer Nature Switzerland AG 2019

N. Diawara (ed.), *Modern Statistical Methods for Spatial and Multivariate Data,*
STEAM-H: Science, Technology, Engineering, Agriculture, Mathematics & Health,
https://doi.org/10.1007/978-3-030-11431-2_9

Indeed, modeling disease spatio-temporal spread using ordinary least squares or weighted least squares will lead to much bias in the estimates and violation of model assumptions (Pace and LeSage 2009). A simple global Moran's value cannot explain area differences, temporal characteristics, or the spatio-temporal relationship of some natural ubiquitous phenomena. Data may show local trends that are not shared globally. This has led many authors such as Anselin (1995) to partition the global spatial domain and compute Moran's index for each subdomain, leading to local indices of spatial association (LISA). However, the impact of time was not incorporated. Spatio-temporal autocorrelation was first introduced by Cliff and Ord (1975), and the concept has been explored by many others over the years (see, for example, Martin and Oeppen 1975; Wang and He 2007; Lee and Li 2017). By adding a time component to the Moran's statistics, the spatio-temporal characteristics lead to better understanding of the spread of the phenomena or the spatial variations.

We propose to evaluate spatio-temporal model association, capturing Moran's statistical spatial autocorrelation values over iterative time under a Poisson process. The model we propose is capable of handling spatial variation and temporal stochastic/dynamic association. Implementation is provided under a multilevel Poisson process that is time dependent, and the dependence among the levels is captured spatially with a rate of increase of the phenomena over time. The rate of increase is built under incrementing time scales for each subarea. The scale parameter effectively creates a region of dynamic model phenomena occurrence. Using Markov chain Monte Carlo techniques, simulated data are generated to investigate Moran's statistics. Diagnostics and comparison of the Moran's statistics are performed. Clusters are built and local features are presented to enhance the understanding of Moran's statistics, trend in time, and functionality of model analysis.

2 Spatio-Temporal Process

For a measurable space $S = \mathbb{R}^d$, a domain $D \subset S$, and a collection of random points $x_1, x_2, \ldots$, let $N(D)$ represent the count or number of points x_i that capture location occurrence of some disease/event in the domain D. Then $N(D)$ can be seen as a measure and is called a point process. For nonoverlapping areas $D_1, D_2, \ldots$, $N(D_1), N(D_2), \ldots$ are mutually independent and for each area D, $N(D)$ follows a Poisson distribution with mean λ_D, where λ_D is the intensity of the process. As a process, a counting may be associated with it and is defined as:

$$N(D) = \sum_{i=1}^{\infty} \mathbf{1}_D(x_i), \text{ where } \mathbf{1}_D(x_i) = \begin{cases} 1 \text{ if } x_i \in D, \\ 0 \text{ otherwise.} \end{cases} \tag{1}$$

To every domain, we consider an associated intensity measure λ_D as $\lambda_D = E(N(D))$, the expected number of points that are obtained in D. Such measure can be assumed to be finite if $N(D) < \infty$, hence the name finite measure.

We consider the temporal conditional distribution of the occurrence of disease in the partitioned subareas defined by Voronoi cells induced by a set of points (sites) x_i^t as described in Kallenberg (2001). These cells are defined as:

$$D(x_i^t) = \{x \in D : \mu(x_i^t, x) < \mu(x_i^t, z), \forall z \in D, \mu \in \mathcal{N}(\mathbb{R}^d)\},$$

where $\mathcal{N}(\mathbb{R}^d)$ is the class of locally finite measures on $\mathbb{R}_+$ and D is the domain area. Hence $D - i^t$ denotes the area around the point x_i^t. Doing so, we have a local precise measurement of the density and neighborhoods. We then build a partition of D into subareas called time dependent Voronoi collection of D, D_i^t where a finite measure or "locally finite" measure is such that $\mu D_i^t < \infty$ for all $i \geq 1$.

Later, we consider as measure the *nonhomogeneous Poisson point process* with non-constant intensity function λ_D in time say, $\lambda_D(t)$ the expected number of points in D in interval time of length t, with $t \in [0, T]$ and the main assumption that points are independent of each other. The nonhomogeneous Poisson process has the following three important properties as described in Baddeley et al. (2016):

- **conditional property:** given exactly n points in a region D_k^t, these points are mutually independent and each point has the same probability distribution over D, with probability density $Y_{k(t)} = Y_{D_k^t} \sim f_{D_k}(t) = \lambda_{D_k}(t)/\mu_D$, where $\mu_D = \int_D \lambda_D(t)dt.$
- **superposition property:** If $Y_1, Y_2, \ldots, Y_n$ are independent Poisson random variables with means $E(Y_i) = \lambda_i$, $\lambda_i \in \mathbb{R}^+$, then $\sum_i Y_i \sim Poisson(\sum \lambda_i)$.
- **random thinning property:** Suppose that $N \sim Poisson(\lambda)$, and that $Y_1, Y_2, \ldots, Y_n$ are independent, identically distributed multinomial random variables with distribution $Multinomial\ (p_1, p_2, \ldots, p_n)$, that is $P\{Y_i = k\} = p_k$ for $k = 1, 2, \ldots, m$. Then the random variables $N_1, N_2, \ldots, N_m$ defined by
$$N_k = \sum_{i=1}^k \mathbf{1}\{Y_i = k\}$$ are independent Poisson random variables with parameters $E(N_k) = (\lambda p_k)$.

Hence $\lambda_{D_k^t} = \lambda(\cdot, D_k^t)$ is a random variable in $[0, +\infty]$ for every $D_k^t \subseteq D$. Such an idea is also described in Thäle and Yukich (2016) with properties of Poisson-Voronoi generated sets.

The Voronoi cells are defined as a point process $D_k^t \subseteq D$, each D_k^t containing a finite number of simple points and they can be viewed as a countable set of random points with associated intensity $\lambda_D = E(N(D_i))$ the expected number of points in D_i.

The cells D_k^t define a sequence of original/congruent points and for partition $\{D_k^t\}$ of D, each D_k^t can be associated with the original point x_k^t. If $D(t)$ denotes the collection of Voronoi cells generated at time t, $D(t) = \bigcup_{k(t)=1}^{n(t)} D_{k(t)}^t$, the elements of $D(t)$ are nested within $D(t-1)$, i.e., $D(t)$ is a refinement of $D(t-1)$. They include convex sets (Kieu et al. 2013), differentiable manifolds with smooth boundary of finite dimensional Hausdorff-type measure with topology on $\mathbb{R}$.

We formalize the notion of conditional distribution of the count given that a point has occurred under Palm probability/measure and define it as $\mathbb{P}_{x_{k(t)}^t}(D_k^t)$ and because of time dependence we consider the reduced Palm distribution $\mathbb{P}^!_{x_{k(t)}^t}(D_{k(t)}^t)$ denoted as $\mathbb{Q}_{x_{k(t)}^t}$, which is the conditional distribution of the count when the point at x is omitted.

The cells are such that their asymptotic mean and variance are a function of the weighted surface (Reitzner et al. 2012).

For a random exchangeable sequence of spaces $(D_1^t, \ldots, D_{n_t}^t)$ defined at time $t \in [t_{i-1}, t_i)$, let $\mathbb{Q}_{k(t)}^t$ denote the empirical distribution of the sets generated from points $x_{k(t)}^t$, $1 < k(t) < n(t)$. Let $(Y_{k(t)})$ be the sequence count of outcomes observed in $(D_{k(t)}^t)$, then

$$P\left(Y_1, \ldots, Y_{n_t}\right) = \int \prod \mathbb{Q}_{x_i^t}, \text{ or}$$

$$P\left(Y_{k(t)}, \ 1 \le k(t) \le n(t)\right) = \int \mathbb{Q}_{x(t-1)}(D_{k(t)}) dF(\mathbb{Q}),$$

where $\mathbb{Q}$ is the limiting distribution of F_N over D.

Observing that the sequence is time dependent, we introduce the model as a discrete time Markov chain as in Resnick (2002):

$$P\left(Y_{k(t)}|Y_{k(1)}, \ldots, Y_{k(t-1)}\right) = P\left(Y_{k(t)}|Y_{k(t-1)}\right)$$

and define the process as follows:

$$P\left(Y_1^t, \ldots, Y_n^t|Y^{t-1}\right) = \int \mathbb{Q}_{x_1}(D_1^t) \cdots \mathbb{Q}_{x_n(t)}(D_n^t) dF(\mathbb{Q}), \text{ or}$$

$$P\left[\left(Y_{k(t)}\right)_{1 \le k(t) \le n(t)} |Y_{k(t-1)}\right] = \int \prod \mathbb{Q}_{x_{k(t)}}(D_k^t(t)) dF(\mathbb{Q}).$$

For a time period of length T partitioned into m time subintervals $t_0 = 0 < t_1 < t_2 < \cdots t_r < \cdots < t_m$, consider $[t_{r-1}, t_r)$, $r = 0, 1, \ldots, m$, $0 \le t_r \le T$. The number of random occurrences or points generated within (t_{r-1}, t_r) in state space D are tied to locations which are sequences of nested counts of total number of points generated at time t_r, denoted $n(t_r)$, and sequence of points $k(t_1), k(t_2), \ldots, k(t_m)$, where $1 \leq k(t_1) \leq n(t_1); 1 \leq k(t_2) \leq n(t_2); \ldots; 1 \leq k(t_r) \leq n(t_r); \ldots; 1 \leq k(t_m) \leq n(t_m)$, and $n(t_r)$ represents the total number of points generated at time t_r within specific locations. The mathematics then requires notation capable of representing outcomes across multiple time point steps and subareas. The associated count within D_k^t defines a sequence $\{Y^{t_r}\}$ of number of occurrences within $[t_{r-1}, t_r)$ in subarea $D_{k(t)}^t$ and follows the Markov property, i.e., given the current state of the system, we can make predictions about the future state without regard for previous states. This is a discrete time finite Markov chain.

To minimize complexity, we define the count process $\{Y^t, t \geq 0\}$ of occurrences between consecutive times t_{r-1} and t_r to be

$$P_{Y^t}(t_r - t_{r-1}) = P(\lambda(t)_D|D|),$$

where $\lambda_D(t)$ denotes the local intensity for any t within $[t_{r-1}, t_r)$. Such a process has stationary transition probabilities and transition matrix of the counts $Q = (Q_{ij})$ constructed from the transition probabilities, $P_{ij}(s)$, where Q_{ij} is the (i, j) count from $P_{ij}(s)$ from consecutive time periods $t_1, t_2, \ldots, t_m$ adjusted with subarea containing locations i and j, and $Q_{ii} = 0$, $\forall\, i \in S$, at each time period.

We extend D_k^t such that they form a nested sequence $D_{k(t_r)}^{t_r} \subset D_{k(t_{r-1})}^{t_{r-1}}$, the subarea within time (t_{r-1}, t_r) and count $Y_{k(t)}^t$, $1 \leq k(t) \leq n(t)$. Following Resnick (2002), we can define the nested subareas $D_{k(t)}^t$ and each fixed time t, a sequence $\{E_{k(t)}^t\}_{k \geq 0}$ of independent and identically distributed exponential random variables with unit mean such that $E_{k(t)}^t$ is independent of $Y(t)$ on $D_{k(t)}^t$, $1 \leq k(t) \leq n(t)$ and set $Y_{k(t)}^t$ equal to the count within $D_{k(t)}^t$ for interval time $[t_{r-1}, t_r)$.

By discretizing time, we then define two finite sequences $\{(Y^{t_r}), (t_r)\}$, where

$$Y^t = \sum_{k(t)=1}^{n(t)} Y_{k(t)}^t \text{ with locations } \underset{\sim}{Y}_{k(t)}^t \text{ for } t_n < t < t_{n+1} \text{ and } \{t_r\}, \; r = 1, \ldots, m$$

is the sequence of times when the process is observed. The sequence $\{Y^t\}$ is the cumulative count of occurrence within interval time $[t_{r-1}, t_r)$ in a subset of space D such that $t_r - t_{r-1}$ is conditionally independent and exponential given Y^{t_r-1}. More precisely, the sequence $\{Y_{k(t)}^t\}$ is defined as follows:

$$Y_1^{t_1}$$

$$\vdots$$

$$Y_{k(t_1)}^{t_1}$$

$$\vdots$$

$$Y_{n(t_1)}^{t_1}$$

$$\text{and } Y^{t_1} = \sum_{k(t_1)=1}^{n(t_1)} Y^{t_1}_{k(t_1)} \, ;$$

$$\vdots$$

$$Y^{t_r}_1$$

$$\vdots$$

$$Y^{t_r}_{k(t_r)}$$

$$\vdots$$

$$Y^{t_r}_{n(t_r)}$$

$$\text{and } Y^{t_r} = \sum_{k(t_r)=1}^{n(t_r)} Y^{t_r}_{k(t_r)} \, ;$$

$$\vdots$$

$$Y^{t_m}_1$$

$$\vdots$$

$$Y^{t_m}_{n(t_m)}$$

$$\text{and } Y^{t_m} = \sum_{k(t_m)=1}^{n(t_m)} Y^{t_m}_{k(t_m)},$$

where $n(t_r)$ is the total number of points generated at time t_r.

3 Moran's Statistics

The global Moran's index, I, based on a sample of n observations is defined as:

$$I = \frac{n}{\sum_{i \neq j} w_{ij}} \frac{\sum_{i=1}^{n} \sum_{j=1}^{n} w_{ij} Z_i Z_j}{\sum_{i=1}^{n} Z_i^2},$$

where Z_i and Z_j are the normalized values of spatial characteristic or feature indications i and j, and w_{ij} is the weight between features i and j (see Moran 1950).

Using Eq. (1), Vaillant et al. (2011) modeled the propagation of sugarcane yellow leaf virus with a focus on the spatial spread of disease over six time periods (weeks 6, 10, 14, 19, 23, and 30 in the growing season). For each pair of consecutive observation dates (t_{i-1}, t_i), $i = 1, 2, \ldots, 6$, they defined a Moran-type index based on a nearest neighbor scheme as follows:

$$M_i = \sum_{(x,y) \in D} w_{x,y} \mathbf{1}_{[0,t_{i-1}]}(T_x) \mathbf{1}_{(t_{i-1},t_i]}(T_y),$$

where D denotes the discrete set of plant locations, T_x denotes the date (time variable) of virus detection for plant x, $\mathbf{1}_{[0,t_{i-1}]}(T_x)$ is an indicator whether time T_x falls in the interval $[0, t_{i-1}]$, and $w_{x,y}$ denotes the weights, which are non-zero only if x and y are neighbors, located within the same subarea. There are several possible weight functions as we will describe later in this section.

We will utilize a similar space-time definition of Moran's index. Since the partitioned areas vary from one time to the next, we define D_k^t as subarea $k \geq 1$ at time t and then we will define a Moran-type autocorrelation statistic on each measurable subset D_k^t as:

$$M_k^t = \sum_{u,u' \in D_k^t} w_{u,u'} \mathbf{1}_{[t-1,t)}(T_u, T_{u'}), \quad k \geq 1, \quad t \geq 1, \tag{2}$$

where $w_{u,u'}$ denotes the spatial weight between points u and u' of disease/event occurrence ($w_{u,u} = 0$), T_u and $T_{u'}$ denote the time of detection of u and u', and $\mathbf{1}_{[t-1,t)}(T_u, T_{u'})$ is an indicator of whether times T_u and $T_{u'}$ both fall in the interval $[t-1, t)$.

In the definition of Moran's autocorrelation index, $w_{u,u'}$ represents a spatial weight between any two distinct event locations generated within disc D_k^t, which could be a function, e.g.,

 (i) the inverse distance between two points;
 (ii) the inverse distance squared between two points;
(iii) an estimate of the autocorrelation/semivariance statistic;
(iv) the "geographical" weights defined as

$$w_{ij} = \begin{cases} e^{(-d_{ij}/r)} & j \neq i, \\ 0 & \text{otherwise}, \end{cases}$$

where r is the maximum distance in the minimum tree that spans all points and does not have any nodes that link back to itself as in Murakami et al. (2017).

As stated in Chen (2012), selection of the weight function objectively is an open question. Different weight functions have different spatial effects. Later in Sect. 4, we give reasons why we choose a particular weight function.

Also, while values of the Moran-type statistic may not always be between -1 and 1 (the full range depends on the weights), they are essentially in the same spirit as larger values indicate larger autocorrelation measures.

We will use a modified version of Geary's C autocorrelation index (Geary 1954; Sokal et al. 1998) as a comparison to determine and evaluate the trend between the two measures. We will define it as C_k^t:

$$C_k^t = \frac{n-1}{2W_0} \sum_{u,u' \in D_k^t} w_{u,u'} \mathbf{1}_{(0,t-1]}(T_u) \mathbf{1}_{(t-1,t]}(T_{u'}),$$

with $w_{u,u'}$ as in Eq. (2) and $W_0 = \sum_{u,u' \in D^t} w_{u,u'}$, the sum of weights for all points generated across all discs at time t.

3.1 Expected Value

Cliff and Ord (1981) derived the moments of Moran's index given the assumption that observations are random independent drawings from normal populations. Under this assumption, the expected value of Moran's index is given as:

$$E(I) = \frac{n}{S_0} \frac{E\left(\sum_{i,j} w_{ij} Z_{u(t)} Z_{u'(t)}\right)}{E\left(\sum_i Z_i^2\right)} = -\frac{1}{n-1},$$

where $S_0 = \sum w_{ij}$. To link our modified Moran-type autocorrelation to the global Moran's index, it can be viewed as

$$M_k^t = \sum_{u,u' \in D_k^t} w_{u,u'} \mathbf{1}_{[t-1,t)}(T_u, T_{u'})$$

$$= \sum_{u,u' \in D_k^t} w_{u,u'} Z_{u(t)} Z_{u'(t)},$$

where $Z_{u(t)} = \mathbf{1}_{D_k^t}(T_u) = \begin{cases} 1 & \text{if } u \text{ appeared in } D_k^t \text{ in the time period } [t-1,t), \\ 0 & \text{otherwise.} \end{cases}$

Then $Z_{u(t)}$ is the indication of occurrence of event within the disc centered at an event from the previous time point.

Given the assumption that observations are random independent drawings within the same disc from a given distribution, the expected value of the proposed spatio-temporal Moran's autocorrelation statistic can be derived as follows:

$$\mathrm{E}(M_k^t) = \mathrm{E}\left(\sum_{u,u' \in D_k^t} w_{u,u'}(Z_{u(t)} Z_{u'(t)}) \right)$$

$$= \sum_{u,u' \in D_k^t} w_{u,u'} \, \mathrm{E}\left(Z_{u(t)} Z_{u'(t)} \right)$$

where u and u' are events that occurred under randomization process.

4 Simulation Study

Here we provide an illustration of the theory built in the previous section. We conduct the simulation using R with the `spatstat` package and represent the weights of the Moran's index as the inverse distance between two points. The weight function was chosen as described in Chen (2012) since we are modeling spread of disease which is a large-scale complex system and we intend to capture transitions globally to locally. We begin with an observed unit area (1×1 unit) and generate a Poisson point processes within that area. The Poisson process is used as general setup since it is widely applicable and is applicable to many real-world situations such as disease cases, radioactive decay, and plant propagation. Other processes, for example, negative binomial and Cox, may be utilized to explain other space-time point phenomena. Next, we define the Voronoi cells around each point (e.g., the locus of locations nearer to that point than any other). Each location site is then used as the center point to generate a disc with diameter equal to the minimum of the distance to the Voronoi edge and the distance to the observed area edge (edge of D). This effectively means that an infected site can only subsequently infect sites that are closer to it than any other infected site. These Voronoi discs define the subareas $D_{k(t)}^t$ for the point process generated at the next time interval (step). In application, subareas can be tied to locations such as country, state, county, and zip code. The choice of subareas could be used to determine whether more affluent regions are able to deal more effectively with disease outbreak, for example. The Moran's statistic is calculated at each step as the sum of the inverse distances between an offspring site and another offspring site. This format will continue through time $t = 5$ adhering to the Markov property ($t = 5$ chosen to follow results from a forestry example in Meddens and Hicke 2014 or as in Vaillant et al. 2011). Each subarea will be magnified by a scale parameter, $\alpha > 0$, such that the new radius is α times the previous radius with intensity function $\lambda * t$, following the spatial resolution of aggregated grid cells that experienced a 1% tree mortality as in Meddens and Hicke (2014).

Algorithm: Iterative local Moran's statistics
Procedure:

  (i) Define the partition and generate a Poisson process in the area with intensity λ.
 (ii) Calculate the Moran's statistic.
(iii) Define new subareas with point from previous process as center and radius equal to the minimum of distance to the edge of the window or distance to the edge of the Voronoi cell.
 (iv) Rescale subarea and iterate local points from a Poisson process with intensity $\lambda * t$.
  (v) Compute the local Moran's statistic for each subarea at that time step.

Repeat   steps (iii) through (v).
Stop   when all times are reached.
End

Figure 1 shows the original observed area with the seven points generated at $t = 1$ alongside the discs generated from these points and the points generated at $t = 2$.

The Moran's value for $t = 1$ equals 31.65. The number of points generated in each Voronoi subarea, the rescaled area, and Moran's values generated at time $t = 2$ are displayed in Table 1 with $\lambda = 4$ and $\alpha = 5$. For example, at point 1 in Fig. 2a, we generate a disc of maximum radius within the Voronoi cell centered at the point generated at the previous time with 10 points as shown in Fig. 2b. The number of points generated in other Voronoi cells is displayed in Table 1. The Geary's values are also computed. The ordering of Moran's and Geary's values display resemblance as they "rank" the strength of correlations in a similar way.

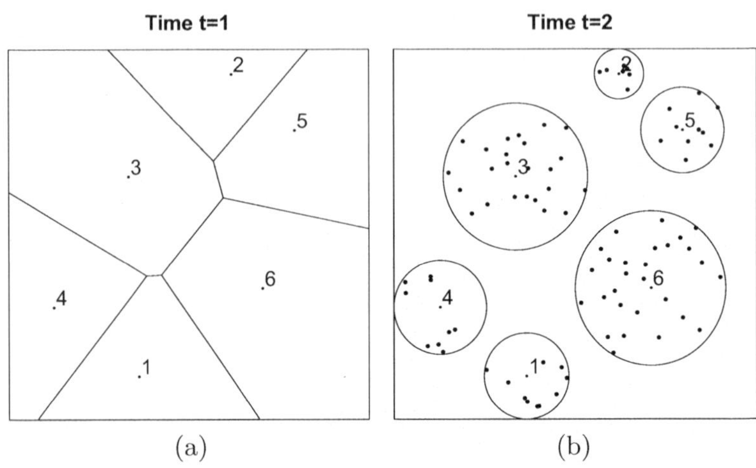

Fig. 1 Poisson point process plots, $\lambda = 4$ **(a)** $t = 1$ and **(b)** $t = 2$

Table 1 Non-zero Moran's and Geary's c values for $\lambda = 4$ and $\alpha = 5$ at $t = 2$

Disc	Points	Area	Moran's	Geary's C
1	10	1.064	158.18	2.444
2	7	0.361	129.29	1.998
3	24	3.096	453.09	7.000
4	9	1.266	111.58	1.724
5	10	1.058	116.03	1.793
6	28	3.415	542.49	8.382

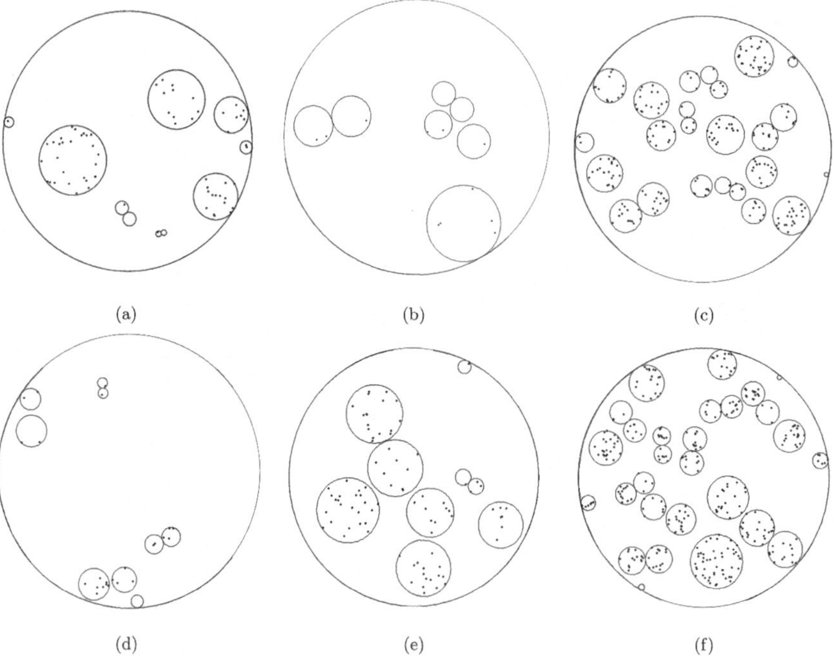

Fig. 2 Discs 1–6 (**a–f**) at $t = 3$

The process is described at time $t = 3$ with disc 1 having 10 subareas (see Fig. 3), disc 2 having 7 subareas, etc. The non-zero Moran's values for disc 1 at time $t = 3$ are listed in Table 2.

After five time intervals, the clustering of points generated is displayed in Fig. 3, with red shades representing the most dense areas and white shades representing areas with no instances of disease. Such a representation would not have been clearly observed if the area was not partitioned and if the rate of spread was not tabulated in space and time. This framework will allow us to detect clusters and estimate the density.

Fig. 3 Apparent clustering
through $t = 5$

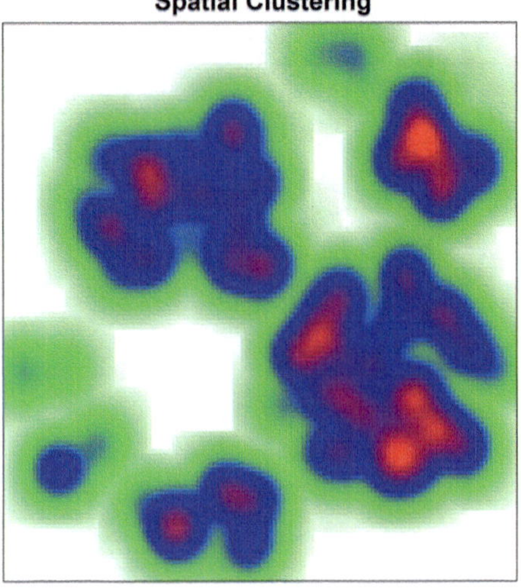

Table 2 Non-zero Moran's
and Geary's c values for disc
1 at $t = 3$

Disc	Points	Area	Moran's	Geary's C
1	28	1.905	779.89	5.101
5	11	1.362	125.23	0.819
8	2	0.065	20.11	0.132
9	15	1.183	337.63	2.208
10	7	0.508	80.63	0.527

5 Model Analysis

Here we investigate any trend in our Moran's values based on the number of points
generated and area of the disc. Figure 4 shows area and number of points plotted
versus Moran's value. The plots of Moran's values for time $t = 2$ and for disc 1
at $t = 3$ and $t = 4$ show an approximate linear trend between Moran's values and
number of points generated, even though there is no apparent relationship between
area and Moran's values.

The 3-D plot in Fig. 5 shows that at each successive time point the number of
points generated versus the Moran's gets progressively larger. This is due to the
number of points generated increasing as time increases, but the area increase is
moderated as time increases. This results in points that are closer together, thus
increasing the sum of inverse distances and the Moran's value.

Focusing on the six points generated at $t = 1$ and the associated discs at $t = 2$,
we investigate the idea of correlation between two consecutive Poisson distributions.
Table 3 shows the number of points generated within each of the original six discs
at times $t = 2, 3, 4, 5$ and also includes the area of each disc. The total number of
points generated is also shown and each of these points has an associated Moran's
statistic at the next time point.

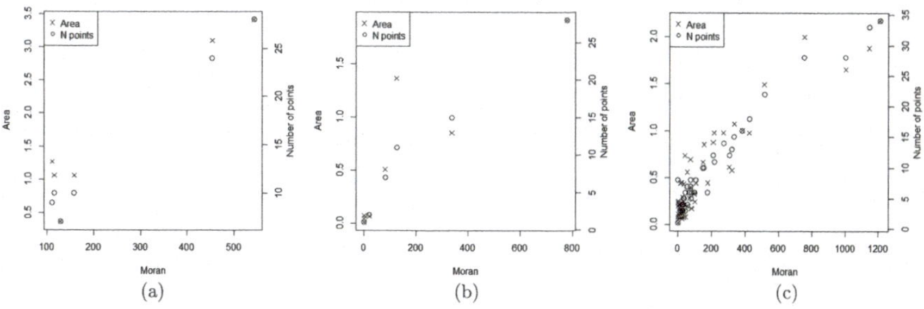

Fig. 4 Trend analysis (**a**) $t = 2$, (**b**) $t = 3$ for disc 1, and (**c**) $t = 4$ for disc 1

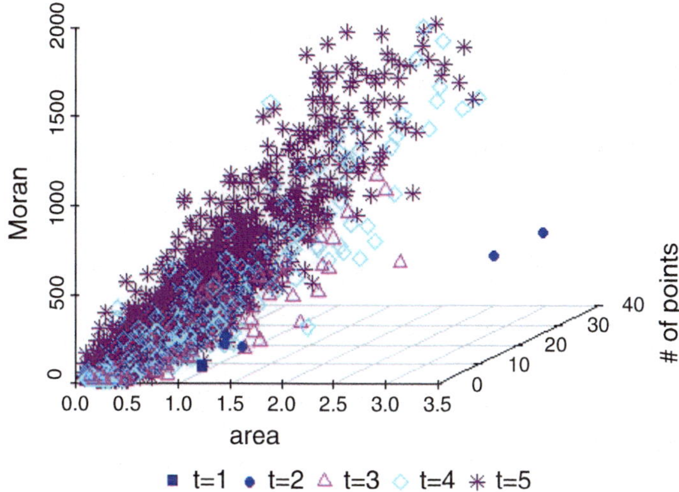

Fig. 5 Plot of Moran's values

Table 3 Counts at $t = 2, 3, 4, 5$

Disc	Area	Time 2	3	4	5
1	1.064	10	66	437	2781
2	0.361	7	11	37	174
3	3.096	24	235	1303	7710
4	1.266	9	24	90	547
5	1.058	10	89	738	4399
6	3.415	28	336	2223	13305
Total points		88	761	4828	28,916

6 Conclusion

Our designed model has shown that global properties of spatio-temporal disease spread can be tuned in by modifying the space and adding time efficiently. The results show that long range, spread feature, and variability yielded spatial clustering of occurrence events. To obtain a quantitative analysis of the performance of the Moran's values, we compared them with adjusted Geary's C statistics. We controlled for targeted subareas and such an approach provides a framework for understanding "disorganized" or "disordered" evolution of some natural organisms in the form of cluster analysis. The distribution of subareas can be found based on our density estimation or concentration under the proposed spatio-temporal Moran's index.

Extension to different dynamics other than the Poisson will be considered in future studies and widen the scope of the spatio-temporal model incorporating covariates.

Acknowledgements The research for this paper was made possible by financial support from the U.S. Naval Academy. Material contained herein is solely the responsibility of the authors and is made available for the purpose of peer review and discussion. Its contents do not necessarily reflect the views of the Department of the Navy or the Department of Defense.

References

Anselin, L.: Local indicators of spatial association - LISA. Geogr. Anal. **27**(2), 93–115 (1995). https://doi.org/10.1111/j.1538-4632.1995.tb00338.x

Baddeley, A., Rubak, E., Turner, R.: Spatial Point Patterns: Methodology and Applications with R. CRC Press, Boca Raton (2016)

Chen, Y.: On the four types of weight functions for spatial contiguity matrix. Lett. Spat. Resour. Sci. **5**, 65–72 (2012). https://doi.org/10.1007/s12076-011-0076-6

Cliff, A., Ord, J.: Space-time modelling with an application to regional forecasting. Trans. Inst. Br. Geogr. **64**, 119–128 (1975). https://doi.org/10.2307/621469

Cliff, A., Ord, J.: Spatial Processes - Models and Applications. Pion Limited, London (1981)

Geary, R.C.: The contiguity ratio & statistical mapping. Inc. Stat. **5**(3), 115–145 (1954). https://doi.org/10.2307/2986645

Jones-Todd, C.M., Swallow, B., Illian, J.B., Toms, M.: A spatio-temporal multispecies model of a semicontinuous response. J. R. Stat. Soc. Ser. C **67**(3), 705–722 (2018). https://doi.org/10.1111/rssc.12250

Kallenberg, O.: Foundations of Modern Probability, 2nd ed. Springer, New York (2001)

Kieu, K., Adamczyk-Chauvat, K., Monod, H., Stoica, R.S.: A completely random T-tessellation model and Gibbsian extensions. Spat. Stat. **6**, 118–138 (2013). http://dx.doi.org/10.1016/j.spasta.2013.09.003

Lawson, A.: Bayesian Disease Mapping: Hierarchical Modeling in Spatial Epidemiology. Chapman & Hall/Crc Interdisciplinary Statistics Series. CRC Press, Boca Raton (2009)

Lee, J., Li, S.: Extending Moran's index for measuring spatiotemporal clustering of geographic events. Geogr. Anal. **49**, 36–57 (2017). https://doi.org/10.1111/gean.12106

Martin, R.L., Oeppen, J.E.: The identification of regional forecasting models using space:time correlation functions. Trans. Inst. Br. Geogr. **66**, 95–118 (1975). https://doi.org/10.2307/621623

Meddens, A.J.H., Hicke, J.A.: Spatial and temporal patterns of Landsat-based detection of tree mortality caused by a mountain pine beetle outbreak in Colorado, USA. For. Ecol. Manag. **322**, 78–88 (2014). https://doi.org/10.1016/j.foreco.2014.02.037

Moran, P.A.P.: Notes on continuous stochastic phenomena. Biometrika **37**(1–2), 17–23 (1950). https://doi.org/10.1093/biomet/37.1-2.17

Murakami, D., Yoshida, T., Seay, H., Griffith, D.A., Yamagata, Y.: A Moran coefficient-based mixed effects approach to investigate spatially varying relationships. Spat. Stat. **19**, 68–69 (2017). https://doi.org/10.1016/j.spasta.2016.12.001

Pace, R.K., LeSage, J.P.: Omitted variable and spatially dependent variables. In: Páez, A., et al., (eds.) Progress in Spatial Analysis: Advances in Spatial Science, pp. 17–28. Springer, Berlin (2009)

Reitzner, M., Spodarev, E., Zaporozhets, D.: Set reconstruction by Voronoi cells. Adv. Appl. Probab. **44**(4), 938–953 (2012). https://doi.org/10.1239/aap/1354716584

Resnick, S.I.: Adventures in Stochastic Processes. Birkhäuser, Boston (2002). https://doi.org/10.1007/978-1-4612-0387-2

Sokal, R.R., Oden, N.L., Thomson, B.A.: Local spatial autocorrelation in a biological model. Geogr. Anal. **30**(4), 331–354 (1998). https://doi.org/10.1111/j.1538-4632.1998.tb00406.x

Thäle, C., Yukich, J.E.: Asymptotic theory for statistics of Poisson-Voronoi approximation. Bernoulli **22**(4), 2372–2400 (2016). https://doi.org/10.3150/15-BEJ732

Vaillant, J., Puggioni, G., Waller, L.A., Daugrois, J.: A spatio-temporal analysis of the spread of sugarcane yellow leaf virus. J. Time Ser. Anal. **32**, 392–406 (2011). https://doi.org/10.1111/j.1467-9892.2011.00730.x

Wang, Y.F., He, H.L.: Spatial Data Analysis Method. Science Press, Beijing (2007)

Wang, H., Cheng, Q., Zuo, R.: Quantifying the spatial characteristics of geochemical pattern via GIS-based geographically weighted statistics. J. Geochem. Explor. **157**, 110–119 (2015)

Zhou, H., Lawson, A.B.: EWMA smoothing and Bayesian spatial modeling for health surveillance. Stat. Med. **27**, 5907–5928 (2008). https://doi.org/10.1002/sim.3409

Printed by Printforce, the Netherlands